The Mammalian Jaw

The structure of mammalian skulls is notably diverse; however, at a fundamental level the jaw mechanism is remarkably similar, if not essentially the same, in the majority of mammals.

Using simple models that are compared with real animals at every step, this book examines the basic structural features of the mammalian jaw mechanism from a mechanical point of view. It explores how the mechanical constraints placed on the jaw have contributed to the evolution of an efficient underlying structure, used by many mammals, which precludes mechanical difficulties and uses a minimum amount of bone tissue. Throughout the book the emphasis is on conceptual understanding, with explanations linked together to form a complete story that can be applied to both fossil and extant mammals.

Walter Stalker Greaves is a retired member of the Department of Oral Biology, University of Illinois at Chicago. He spent his career teaching anatomy at Indiana University of Pennsylvania and human gross anatomy in the College of Dentistry, UIC. His research concentrated on the mammalian jaw mechanism, using very simple mechanical models to explain how the jaw works and why the jaw mechanism has evolved to essentially the same basic structure.

The Mammalian Jaw

A Mechanical Analysis

WALTER STALKER GREAVES
University of Illinois at Chicago

CAMBRIDGE
UNIVERSITY PRESS

CAMBRIDGE UNIVERSITY PRESS
Cambridge, New York, Melbourne, Madrid, Cape Town,
Singapore, São Paulo, Delhi, Mexico City

Cambridge University Press
The Edinburgh Building, Cambridge CB2 8RU, UK

Published in the United States of America by Cambridge University Press, New York

www.cambridge.org
Information on this title: www.cambridge.org/9781107016224

First published 2012

Printed in the United Kingdom at the University Press, Cambridge

A catalogue record for this publication is available from the British Library

Library of Congress Cataloguing in Publication data
Greaves, Walter Stalker, 1937–
The mammalian jaw : a mechanical analysis / Walter Stalker Greaves.
pages cm
Includes bibliographical references and index.
ISBN 978-1-107-01622-4
1. Mammals–Anatomy. 2. Jaws–Mechanical properties. I. Title.
QL739.G74 2012
599.14′4–dc23
2012013670

ISBN 978-1-107-01622-4 Hardback

For

Marsha Lee and Jacqueline Sinclair

Contents

Preface *page* ix
Acknowledgments xi

Introduction 1

1 **The jaw viewed as a two-dimensional lever** 3

1.1 Location of the muscle force 3
1.2 Moments of force 5
1.3 Reaction forces 6
1.4 Bite force along the tooth row 9
1.5 Joint forces 12
1.6 The length of the tooth row 18
1.7 The location of the muscle vector 19
1.8 Summary 24

2 **The jaw viewed as a three-dimensional lever** 26

2.1 The three-dimensional jaw 26
2.2 The triangle-of-support 27
2.3 The distribution of joint and bite forces 29
2.4 Tensile forces at a jaw joint in the three-dimensional case 31
2.5 Different locations of the resultant force 34
2.6 A typical jaw 37
2.7 A contour map of the bite forces 41
2.8 Location of the carnassial tooth 44
2.9 The mediolateral location of the tooth row 45
2.9.1 The sum of the bite forces 45
2.9.2 The length-to-width ratio of the jaw 46
2.10 Occlusion of the tooth rows 48
2.11 Tooth-wear patterns 50
2.12 Summary 50

3 Vector inclination and joint location 52

3.1 The inclination of the resultant vector of jaw muscle force 52
3.2 Joint and tooth row on the same line 55
3.3 Joint above the tooth row 56
3.4 Posteriorly inclined vectors 59
3.4.1 The direct distance from the joint to the third molar 59
3.4.2 Minimizing a pair of distances 64
3.4.3 Using four points to model the jaw mechanism 67
3.4.4 The spatial configurations of the four points 68
3.4.5 Point T lies on the vector 72
3.5 Anteriorly inclined vectors 74
3.5.1 A relationship between an anterior vector and a long diastema 75
3.5.2 *Glyptodon* 79
3.5.3 Entelodonts 80
3.6 Estimating vector location from muscle weights and the location of the third molar 81
3.7 Summary 84

4 Skull torsion and the postorbital bar 85

4.1 Skull torsion 85
4.2 Selenodont artiodactyls 88
4.3 Torsion in the skull during canine biting 90
4.4 Summary 93

General summary 95

References 101
Index 113

Preface

I initially studied the jaw mechanism in fossil selenodont artiodactyls (antelope and their kin) and augmented the study with an examination of modern forms. One could argue that this was not the best approach, because the jaw mechanism of artiodactyls, while basically similar to other mammals, is in some ways slightly atypical. At the same time, however, some features appear to be somewhat less complicated in the artiodactyls. So it is just as reasonable to imagine that even though these animals were in some ways not the best animals to choose for an *initial* study of the mammalian jaw mechanism, the simplicity of some features made some early progress possible in a reasonable amount of time.

Later studies naturally concentrated on problems that presented themselves at the time. While this is a perfectly reasonable approach, the difficulty is that a clear, logical, and sequential presentation of how the mammalian jaw mechanism works would not necessarily be evident from an examination of a series of papers prepared in this way. The reader, in effect, would be asked to integrate a large amount of unordered information to produce a logical, linear sequence. My view of how the jaw mechanism functions evolved over time as more information came to light, as my early errors were corrected, and as erroneous ideas from the literature were rectified. In short, someone with an interest in the mammalian jaw mechanism might not have an easy time deciphering my views if they simply read my various papers in their order of appearance.

Thus, the purpose here is to present in a clear, logical, and linear sequence, my view of the basic structure of the mammalian jaw mechanism from a geometric point of view. The emphasis is on what this basic structure is and, just as importantly, why, in a mechanical sense, the masticatory apparatus has evolved to the same basic structure. This presentation begins at a very elementary level and, while maintaining as much simplicity as possible, moves along step by step in a linear fashion. For example, the first analysis described below treats the jaw as a line with a joint on one end, the teeth at the other, and a muscle vector somewhere in between. Two reasonable questions could be asked of such a model: where, along the line representing the jaw, is the muscle vector located and how long is the tooth row? Naturally, simpler models must give way to those that are more complicated. The hope is to provide a clear explanation that accounts for the present structure (or natural design) of the mammalian jaw mechanism. Some workers approach studies of this kind by studying the basic physics, say, and then by applying the appropriate equations to apparent problems. Here I find it more satisfying to start with the problem and then try to find an "equation."

In the following account references are largely omitted with the hope of producing a smoother presentation. Readers who prefer a slightly more formal discussion, of course, can consult my original published papers as well as the other papers in the References section. However, they then must integrate the material into a smooth linear sequence for themselves.

Acknowledgments

Over the last 40 years, many students and colleagues have contributed, in both large and small ways, to my ideas about the jaw mechanism. At the beginning, my thesis advisor Len Radinsky and my committee members Jim Hopson, Ralph Johnson, Leigh Van Valen, and especially Charles Oxnard, had a great influence.

For all her efforts I especially want to thank Sue Herring, who was both a fellow student at the University of Chicago and later a colleague at the University of Illinois at Chicago. Very little escaped Sue's sharp mind. Her efforts to teach me to spell, however, were less successful, especially after spell-checkers became common. My other colleagues at Illinois are fondly remembered. Many hours were spent with Herb Barghusen, who challenged me at every turn. While he found errors in my thinking, I remember the look on his face one day after a long discussion that suggested to me that he thought there might (finally) be something to what I was telling him; but he never quite said so. Bob Scapino also constantly challenged me, but I think his greatest contribution to my career was to ensure that I was able to pursue the research I wanted to, while he was head of the Department of Oral Anatomy and later Oral Biology. Karen Hiiemae helped me to crystallize in my mind the different approaches to research. Writing a paper with Ray Costa gave me a better appreciation of other fields of study. Tom Lakars was constantly encouraging and questioned me in a calm and friendly manner. Lloyd DuBrul hired me and made possible these relationships with the functional anatomists in his department.

Graduate and post-doctoral students probably don't realize what a strong effect they have on one's ideas. Discussing things with students certainly clarifies your thinking. I well remember some of my discussions with Fred Anapol, Keith Condon, Doreen Covey, Dean Dessem, Don Dunbar, Bob Druzinsky, Ken Gordon, Bruce Manion, Bob Mucci, Ginny Naples, Al Obrez, Bob Schmitz, Kenshu Shimada, and Larry Wineski. Audrone Biknevicius and Mark Spencer were graduate students from other schools who influenced me and whose research I have since followed carefully.

There is another component of an effort like this. Repeated enhancement of my ideas took place when preparing lectures in gross anatomy that touched on the topics discussed here.

My wife Marsha has illustrated, and questioned me about, every paper that led up to this book. She also greatly improved this manuscript. Of course, any errors or bad ideas are mine.

Finally, I should thank Daryl Domning who, every time he saw me, pressed me to write this book.

Introduction

A number of simple mechanical models are used to analyze the mammalian masticatory apparatus. Beginning with an elementary two-dimensional lever system, additional equally simple models are used to explain basic structural features (the natural design) of the generalized mammalian jaw mechanism. Typically, a single feature of the jaw mechanism is emphasized in each study because its presence in many or most members of a group suggests that such a feature is important functionally. Using a simple model, the best possible configuration for the subsystem that includes the feature is then determined, consistent with the actual conditions in real animals. These models attempt to capture major features of the jaw mechanism rather than the finer details.

More details emerge as additional simple models are considered, but the main features remain paramount. Two major results of studies like this are that: (1) almost all mammalian jaw mechanisms have essentially the same *basic* structure; and (2) this approach has provided a general mechanical explanation for this basic structure. The finding that basic parts of the jaw mechanism are in some sense optimized makes it virtually inevitable that the most basic masticatory apparatus, found in the majority of mammalian groups, will be very similar. One probable reason that such similarities in the skulls and jaws of mammals are not more obvious is that the more superficial structural diversity of mammalian skulls and jaws is impressive and tends to mask the similarities in the more basic features.

Clearly, the overall structure of the mammalian jaw mechanism is extremely diverse. Best known, perhaps, is the diversity of the teeth. Although tooth loss is very common in mammals, the different kinds of teeth are positioned in essentially the same relative locations along the jaw.

The range in the shapes of the jaw itself is also noteworthy, but a basic structure is evident at the outset: joint at the rear, teeth at the front, and muscles in between.

Although their size and orientation are different in different animals, the same large muscles close the jaw. Moreover, they generally attach at equivalent places on the skull and jaw. In a few cases essentially a single muscle mass closes the jaw. Here, distinct muscles are difficult to discern.

The upper part of the jaw joint on the skull, the glenoid, is typically located above the general level of the tooth row. Only in a few cases, such as the fossil entelodonts, is the glenoid cavity on approximately the same level as the tooth row. Some saber-toothed cats may have had the glenoid below the row of teeth, but a number of different straight lines could theoretically be drawn from tooth to tooth so that a particular tooth "row" is difficult to determine. The condylar part of the jaw joint, on the lower jaw, is usually much more variable in position, and although often positioned some distance above the teeth it is just as often close to the same level as the tooth row.

The lower jaw of most mammals is a bar of bone that is often L-shaped, with the jaw joint on one end. The teeth fill a region at the other end of the jaw, and the resultant muscle force provided by the jaw muscles is located somewhere in between. This location of the muscle force in front of, or anterior to, the jaw joint has some significance. In this location the jaw closes when the muscles contract (it would open if the muscle force were located behind the joint).

Two other less obvious similarities seem important. First, there is virtually always a region of some length that is devoid of teeth and is located between the most posterior tooth and the jaw joint. Second, the different parts of the large jaw-closing muscles, while their size and action lines differ, tend to be attached more posteriorly along the jaw. Therefore, if the forces exerted by the individual muscles are imagined to be resolved into a single resultant force, this resultant appears to be located at a point within the posterior half, or even within the posterior third, of the jaw. Even in the case of many rodents, in which some *parts* of the muscle mass that closes the jaw attach rather far forward, the idealized *resultant* muscle force is located closer to the rear of the jaw. The almost universal presence of these two conditions in mammals implies a functional importance within the structure of the masticatory apparatus.

Finally, two other extremely important restrictions on the jaw mechanism are accepted for simplicity. First, maximum, rather than lesser, bite forces are assumed throughout because extreme cases are likely to be the most important. Second, the only biting situation considered in these studies is when the cheek teeth are occluded. That is, when the upper and lower teeth are touching, or (if some food is positioned between the teeth) almost touching.

1 The jaw viewed as a two-dimensional lever

An extremely simple and idealized two-dimensional model introduces this study of the natural design of the mammalian masticatory apparatus. The model treats the jaw as a lever with joint forces at one end, the output or bite forces at and near the opposite end, and the input muscle force somewhere in the middle. Results are constantly compared with the condition in real animals. The major result is that all the teeth will be located in front of the line-of-action of the imagined resultant force of the jaw muscles. The next two results are closely related. First, the line-of-action of the imagined resultant muscle force will be located about one-third of the jaw length from the jaw joint. Second, this location maximizes the sum of the bite forces along the entire length of the tooth row.

The location of the line-of-action is critical, because if the idealized moment arm of the muscles is fixed, then the lever system is also fixed. Bite forces then are determined only by the amount of muscle tissue available.

1.1 Location of the muscle force

At the outset, the importance of bite force to an animal is assumed. Certainly jaws perform many functions, but surely applying force to disrupt food items is one of the most important. As a first approximation, the production of a maximum bite force is an attractive idea. By itself however, it fails to explain the structure of the jaws. As will be seen, only *after* considering the forces at the jaw joint can one return to the idea of a maximum bite force; certain potential problems at the joint must first be avoided.

One way to begin an examination of how mammalian jaws work is to simplify the mechanism of the jaw in order to consider it as a two-dimensional structure and view the lower jaw from the side (Fig. 1.1). In this view, the teeth are attached at the anterior upper edge of the long, more-or-less horizontal, part of the jaw. Behind the teeth, it is usually possible to see three distinct regions of the jaw. First, behind and above the most posterior tooth, a pointed, and often curved, coronoid process serves as much of the site of attachment for the **temporalis** muscle, one of the three great muscles that close the jaw. Behind and below the coronoid is the second region, the articular process. On the end of this process is an articular surface that

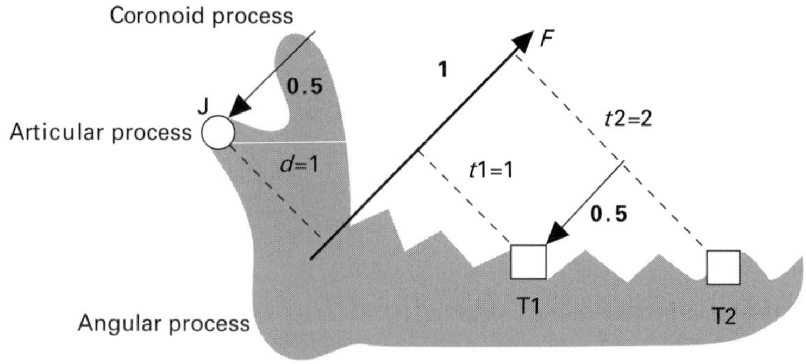

Figure 1.1. The circle (J) represents the jaw joint and the two squares, labeled T1 ad T2, indicate just two of the many teeth in the jaw. The arrow (*F*) is the resultant unit vector of jaw muscle force. The distance from the jaw joint to *F* is indicated by *d*. The distances *t*1 and *t*2 extend from the vector to their respective teeth. Note that all distances are perpendicular to the muscle force vector. The muscle force is 1 unit and the reaction forces (short arrows) at the joint and T1 are each 0.5 units.

forms the lower half of the jaw joint. The angular process (the third region, which in some cases may not be very prominent) forms the rear, lower corner of the jaw and serves as the attachment site for the two remaining large jaw-closing muscles. Much of the **masseter** attaches on the outer or lateral side and part of the **medial pterygoid** attaches to the inner or medial side of this angular process.

In Figure 1.1 the outline of a generalized jaw is indicated in silhouette and only the jaw joint and two representative teeth have been highlighted with symbols (circle and squares, respectively). In addition to the joint and the teeth, the other element needed for the following discussion is the force that is put into the system and that pulls the upper and lower jaws together. Most of this force is provided by the three great jaw-closing muscles already mentioned: temporalis, masseter, and medial pterygoid. The force provided by these three muscles, taken together, is represented by the arrow (**F**) that points up and toward the front of the jaw in this example (Fig. 1.1).

Thus, the arrow, or the resultant unit vector of jaw muscle force stands for the sum of all the forces provided by the many different parts of each of these three large muscles and is imagined to be applied at a single point. The various parts of the muscles actually pull in a number of different directions, but the arrow is meant to represent the orientation of a single force that results when all of the forces, from all of the different parts of the muscles, are considered together. Stated in another way, imagine that all the jaw-elevating muscles are removed. Then, a single force, pulling in the direction of the arrow, would have the same effect as the entire muscle mass where all the parts were active. Of course, Figure 1.1 is a general example. To represent accurately the direction of the resultant in each different mammalian species, theoretically, would require a different drawing for each case. The remaining items in this figure will be discussed at length in the following sections.

1.2 Moments of force

A two-dimensional lever can be represented as a simple line that can rotate around a single point when it is pulled by forces. As the forces provided by the jaw muscles rotate the "bar" of the lower jaw around the jaw joint, the lower jaw, in lateral view, is easily viewed as a lever. Imagining the jaw as such a lever, although far too simple, directs our attention to some important relationships.

Most of us are intuitively familiar with the fact that the magnitude of a force is not the only thing that counts when rotating a structure around a pivot. Push a door at the doorknob, for example, and the door swings easily. Apply exactly the same force close to the hinge and the same door may not move at all. Or put differently, a door can be swung with a small force at, or near, the knob but it takes a larger force to move the same door, with the same ease, if it is pushed farther from the knob and closer to the hinge. Thus, in addition to the force that is applied, the other important variable is its distance from the point of rotation, or the fulcrum. Likewise in a jaw, the magnitude of the muscle force is only part of the story; the effective location of an idealized resultant force is also important.

A single number can be derived from the *magnitude* of a force and its *distance* from a fulcrum. Simply multiplying these two quantities together produces a product called the *moment* of the force. This product, or moment, is an expression of the strength of the tendency to rotate a lever around its fulcrum. If distance is measured in feet and force in pounds, then the moment is expressed in foot-pounds. In the examples below, specific units are dispensed with for the sake of simplicity.

Very often more than one force will act on a lever. If, say, two forces tend to rotate a lever in the same direction, two moments, one for each force, are involved and they simply are added together. If one force tends to rotate the lever in one direction while the other force rotates it in the opposite direction, two possibilities exist. In the first case, the smaller moment is subtracted from the larger. The lever rotates in the direction appropriate for the larger moment, but the effect is reduced by an amount equivalent to the smaller, opposite moment. The second case is the more interesting. Here the opposite moments are equal, and so the lever does not rotate in either direction. The lever is then said to be in equilibrium.

If the *moments* are equal (i.e. the lever is in equilibrium), the *forces* involved can be either equal or unequal. Obviously, equal forces require equal distances from the forces to the fulcrum, if the products of force and distance are to be equal. But if the forces are unequal, the distances from the forces to the fulcrum must also be unequal. Here, the larger force must be paired with the shorter distance and the smaller force with the longer distance in order for the moments to be equal. Thus, a small force can balance a large force if the large force is closer to the fulcrum and the small force is farther away. The products of force and distance ($F \times d$), that is the moments, rotating the lever in opposite directions, are equal in an equilibrium condition ($F_1 \times d_1 = F_2 \times d_2$), and the lever will not rotate in either direction around its fulcrum. An important corollary to note is that the *ratio* of the two forces will be equal to the *ratio* of the two distances, that is $F_1{:}F_2 = d_2{:}d_1$. In order to see clearly how these observations will clarify the workings of the jaw, Figure 1.1 can be examined in more detail.

In Figure 1.1, the circle labeled **J** represents the jaw joint and the two squares labeled **T1** and **T2** indicate just two of the many teeth in the jaw. The arrow (***F***) indicates the line-of-action of the muscle force, and its length indicates a force of one unit of magnitude. The arrow is thus a unit vector. The distance from the joint to ***F*** is indicated by ***d***. The distances ***t*1** and ***t*2** extend from the unit vector to the respective teeth. Note that all distances, as in levers in general, are perpendicular to the unit vector that represents the jaw muscle force.

All the individual jaw muscle forces taken together are expressed as a single resultant force. The moment of force closing the jaw is the product of the magnitude of this muscle force and its distance from the jaw joint, which serves as the fulcrum of the jaw. In Figure 1.1, this product is obtained by multiplying the magnitude of the muscle force (***F***) by the distance ***d***, which is the muscle force vector's distance from the jaw joint. This moment of the muscle force rotates the jaw counterclockwise in this example. The muscle force can be thought of as the input force, because it is the force that is put into the jaw mechanism by the jaw muscles.

In our deliberately simplified example, the distance from the jaw joint to the idealized muscle vector is taken to be ***d***. For ease of calculation, assume that ***d*** is equal to one unit of distance. A muscle force of one unit of force also can be imagined (giving a unit vector), as we are only interested in relative forces. The muscles in a given animal do exert a force with a particular magnitude, but using a number other than 1 complicates the arithmetic without increasing our insight into how the mechanism of the jaw works. To determine the moment of the muscle force, we simply multiply this force (one unit of force) by its distance from the joint (one unit of distance). The counterclockwise moment of muscle force ($F \times d$) is therefore $1 \times 1 = 1$ (again, ignoring what the units of force and distance may actually be in any specific example). How the muscle force compares with the forces at the teeth will be examined next.

1.3 Reaction forces

As the inferred muscle force (***F***) pulls the lower jaw up, the front of the jaw will eventually cease to move any farther because either the upper and lower teeth come into contact or food that is between the teeth is compressed to its limit. The back end of the jaw also tends to move up. The rear of the jaw does not actually move very far because the joint surface is essentially in contact with a fibro-cartilaginous disk that is itself in contact with the joint surface on the part of the skull that forms the upper part of the jaw joint. Thus, there are two *downward* reaction forces that resist the *upward* motion of the jaw. One is at some point along the tooth row, the bite point, and the other is at the jaw joint. They are "reaction" forces because they are reactions to the muscle force; if there were no muscle force there would be no force at the teeth or the joint. In this overly simplified example they are the only reaction forces considered, and so these two forces taken together must be equal to the muscle force because they exactly balance the muscle force. Thus, the upward pull of the muscles on the lower jaw is balanced by two downward pushes; one is a bite force at a tooth, or a few teeth, and the other is a joint force at the jaw joint.

At this point the *general* location of the muscle force can be considered in more detail. If jaw rotation takes place around the jaw joint, then any resultant muscle force passing behind the joint would open rather than close the jaw. A muscle force passing through the joint will naturally be a distance of zero from the joint. The moment of that force will then be 0 and the force will have no turning effect on the lower jaw. It will simply force the joint surfaces toward each other. In order to close the jaw, the muscle force must be located in front of the joint. It must be located a reasonable distance in front of the joint if the moment of force is to have a sufficiently large turning effect. What exactly that reasonable distance might be will be considered in a later section.

When the jaw muscles rotate the jaw around the jaw joint until upward motion ceases because the lower teeth come into contact with a food item or with the upper teeth, the two reaction forces come into play. Once again, if there were no muscle force there would be no force at the teeth or the joints. As the jaw has stopped moving, it is in equilibrium. A jaw in equilibrium can be conceptualized as a bar that has a moment of muscle force rotating it counterclockwise around a fulcrum and an equal, and therefore balancing, moment of reaction force rotating it clockwise (Fig. 1.1). The upward muscle force that rotates the jaw in a counterclockwise direction is the *input* force. The downward reaction force at the bite point is the *output* force. The force at the joint is not rotating the jaw in either direction and will be considered later.

Assume that all the moments are equal when biting on a particular tooth – **T1**, say. The distance from the joint to the tooth is the sum of the distance from the joint to the idealized muscle force (***d***) plus the distance from the muscle force to the tooth (***t*1**). Imagine that ***t*1** is equal to ***d*** and both equal one unit of distance, so that the total distance from the joint to **T1** is two units, or twice the distance from the joint to the muscle vector. If we assume that the muscle force ***F*** is equal to 1, the counterclockwise moment of *muscle* force is equal to 1 ($1 \times 1 = 1$). Therefore, in order to be in equilibrium the clockwise moment must also be equal to 1. As the distance from the joint to the tooth is 2, the output force at the tooth must be 0.5 because 2×0.5 equals a moment of 1 (Fig. 1.1). Here we can recall the example of the door that was given earlier. In this case the muscle force is twice as large as the tooth, or bite force, but these forces have the same, although opposite, rotational effect because the large muscle force is one unit from the jaw joint while the small tooth force is two units from this joint. The two moments, or products of force and distance, 1×1 and 2×0.5, are both equal to 1.

Tooth **T2** provides an additional example. The distance from the jaw joint to this tooth is three units. This can be seen in Figure 1.1, where **T2** is two units from the muscle vector and therefore three units from the jaw joint (***d* + *t*2**). Recall that distance is always measured perpendicular to the vector. The moment of the *muscle* force remains 1, so the moment of the tooth or bite force must also be 1 when the system remains in equilibrium. Because the distance from the jaw joint to the tooth is three units, the tooth force at **T2**, multiplied by 3, must equal 1. Therefore, the tooth force (or output force) in this case is 0.33.

The magnitude of the muscle force is equal to one unit of force and the bite force at **T1** is equal to 0.5 (Fig. 1.1). The upward (counterclockwise) force is greater than the downward (clockwise) force, but the jaw does not rotate, because the moments are equal. At the same time, the entire jaw does not translate directly up toward the skull either, even though the muscle force pulling up is larger than the bite force pushing down. This is true because there is another force in the system.

There are only two places where the lower jaw contacts the skull in this model: the bite point and the jaw joint. Thus, the other half-unit of reaction force is found at the jaw joint. The jaw is not moving in two senses. First, it is not rotating around the jaw joint, because the moments of force are equal. Second, it is not moving straight up as a unit (translating), because the upward muscle force of one unit is balanced by *two* downward reaction forces of a half unit each: one at the bite point and the other at the jaw joint. Essentially the same can be said when biting at tooth **T2** (Fig. 1.1). Here, the reaction forces are not the same as they were when biting on tooth **T1**. Now they are 0.33 at the bite point and necessarily 0.66 at the jaw joint.

Unlike the bite forces at the teeth, however, the joint force is not doing the useful work of the jaw, such as crushing or slicing food, and is in a sense "wasted effort." One might expect that the joint force would be very small if that were possible; the smaller the amount of force dissipated at the joint, the more muscle force that is being used productively at the teeth. However, in mammals the joint force usually has a significant magnitude. There is an important reason for the presence of this joint force, which will be considered in some detail in a later section.

The idealized muscle force vector in this example is oriented up and toward the front of the jaw (Fig. 1.1). This anterior orientation was chosen simply as an example. In this diagram, there is no special significance to an anteriorly oriented, as opposed to a posteriorly oriented, vector. In Figure 1.2 the vector points up and toward the rear of the jaw. The muscle force is again equal to one unit. The vector's distance from the joint, one unit, is the same as the analogous distance in Figure 1.1. The distances from the jaw joint out to the respective teeth are also the same, and therefore, with a muscle force of one unit, the output tooth forces at **T1** and **T2** are also the same: 0.5 and 0.33 when the system is in equilibrium.

The muscle force is posteriorly oriented in most of the great groups of mammals. It is anteriorly oriented in only a few mammalian groups, although there are a large number of species in these groups. This is an extremely important point that will also be considered in greater detail in Chapter 3.

A solid line connects tooth **T1** and tooth **T2**, and extends to the rear as far as the jaw joint (Fig. 1.2). A perpendicular (***h***) has been dropped from the jaw joint to this line. This exercise clearly demonstrates that the jaw joint is located some distance (***h*** units) above the level of the tooth row. Even though it is often difficult or even

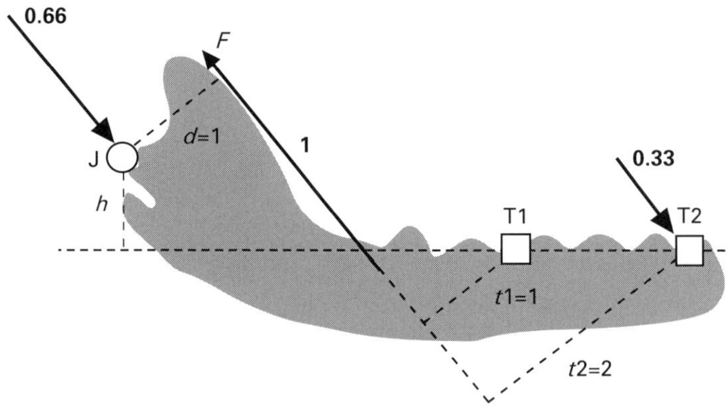

Figure 1.2. This figure represents a jaw with a posteriorly inclined muscle vector rather than one that is anteriorly inclined. The muscle force is 1 unit and the reaction force is 0.66 units at the joint and 0.33 units at the tooth T2. The other labels are as in Figure 1.1.

impossible to draw a straight line through a given tooth row, a joint that is above the general level of the tooth row is typical in mammals. The upper part of the jaw joint (on the skull) is almost always above the level of the tooth row. The lower part of the joint (on the lower jaw) is more variable. The same relationship can also be seen in Figure 1.1. More rarely in mammals, the part of the joint on the skull is situated on, or even below, the level of the tooth row. The significance of this location for the jaw joint, above or below the tooth row, also will be considered in Chapter 3.

In Figures 1.1 and 1.2, the output bite forces at two sample tooth positions were examined. The tooth farther away from the jaw joint exerts a lower output force than the tooth closer to the joint. This is true because as the distance from the joint to the tooth increases, the bite force must decrease if the moment of the bite force is to remain equal to the moment of muscle force.

1.4 Bite force along the tooth row

The previous elementary analyses can be expanded to include a larger number of output bite forces located all along a simple two-dimensional jaw. The same simple calculations that were used above can be used to determine the output force at any number of bite locations along a jaw (bite force is equal to the distance, say 3, from the joint to the vector multiplied by the muscle force of one unit divided by the distance from the joint to the bite point, that is $y = 3/x$). These forces can then be plotted on a graph to give a visual impression of output bite force at different locations along a potential tooth row (Fig. 1.3). On such a graph, bite force is indicated on the vertical axis. Locations along the jaw are marked on the horizontal axis, from the jaw joint, at the left at 0, all the way out to the right-hand side at location 10, where the anterior incisor tooth is located. To simplify the following examples as much as possible, the jaw will be represented as a straight, horizontal line. The

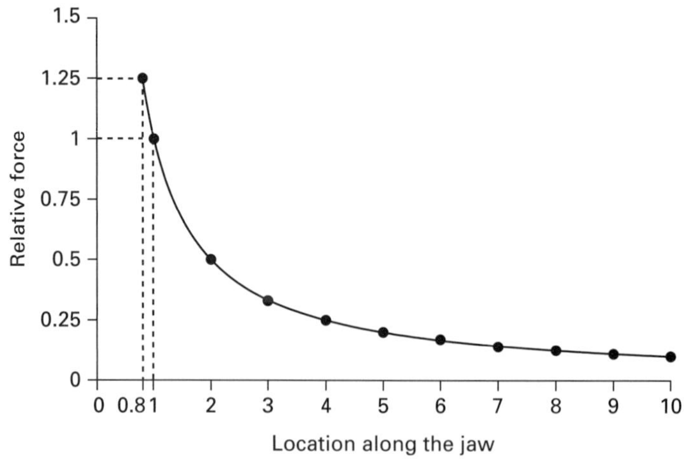

Figure 1.3. A plot of relative bite force against location along a jaw ten units long. The jaw joint is positioned at location 0 and the 1 unit of muscle force is positioned at location 1 in this example.

joint is located at one end, the teeth fill a region of variable length at the other end, and the muscle force is located somewhere in the middle. As another simplification, the resultant muscle force is oriented at right angles to the line representing the jaw; in fact it would usually have a posterior or an anterior inclination in a real animal. Recall that the vector of muscle force must be located at least some distance in front of the joint if contraction of the muscles is to close the jaw. If the vector of muscle force were located behind the jaw joint, the jaw would open, rather than close, when the muscles shortened. If the muscle force vector passed directly through the jaw joint, it would neither open nor close the jaw.

The actual location of the vector of jaw muscle force has only been estimated and has yet to be measured with a great deal of accuracy in a real animal, although a later theoretical analysis, in section 3.6 in Chapter 3, attempts to predict this location. In any event, to see the effect of the idealized location of the muscle force along the jaw, a hypothetical location for a unit vector, say, location 1 can be chosen for consideration. As already mentioned, calculated bite forces at locations closer to the jaw joint are higher than forces at locations that are farther away from the joint. The graph is a smooth curve that slopes down toward the anterior end of the jaw at the right (Fig. 1.3).

In the particular example in Figure 1.3 a force of one unit is assumed for the muscle vector because we are, again, interested only in relative bite forces. Moreover, the assumption is that the muscle vector is applied at location 1. Therefore, the tooth that is also at location 1 will exert a bite force of one unit. In addition, at all potential bite locations that are anterior to the muscle vector and thus farther from the joint, the bite force is lower than the muscle force. This is necessarily true because longer distances to a tooth require correspondingly lower bite forces if a constant moment is to be maintained (recall that an equilibrium condition is assumed). Bite locations that are behind the muscle force, and therefore closer to the joint, are higher than the muscle force, because shorter distances to a tooth require higher bite forces to maintain a constant moment to balance the postulated constant moment of muscle force.

The plot in Figure 1.3 can be examined in more detail with reference to Figures 1.1 and 1.2. The jaw joint is located at location 0. **T1** is at location 2 and is two units from the joint; here the bite force is 0.5. **T2** is at location 3 and is three units from the joint; here the bite force is therefore 0.33. A bite force closer to the jaw joint can also be considered. For example, the bite force at, say, location 0.8 is 1.25 ($0.8 \times 1.25 = 1$). This position is behind the muscle force, and the bite force is higher than the muscle force. As just mentioned, the reason for this is that the moment of muscle force remains the same, so a short distance must be paired with a larger force if the bite force and muscle force moments are to remain equal.

As noted above, generally the jaw muscles are located near the back of the jaw. This seems reasonable from one point of view. If muscles were concentrated at the front of the jaw they could interfere with placing food into the mouth. Moreover, an anterior location for the jaw muscles could limit the amount of mouth opening, or gape. Muscles can stretch only so far. Once they have reached their maximum length, and cannot be stretched any farther, jaw opening ceases. Like the angle between two straight lines, at any amount of jaw opening the distance between the upper and lower jaws is shorter closer to the jaw joint (at the vertex of the angle between the two lines) and longer as it is positioned farther away. Therefore, moving the muscles of a given stretched length farther and farther back allows wider and wider jaw opening. Muscles that are nearer the front of the jaw allow a smaller amount of jaw opening.

From the point of view of bite force production, however, muscles at the front of the jaw seem better and muscles at the rear seem less advantageous. As an example, suppose the muscle vector in Figure 1.3, with one unit of force, were postulated to move one unit toward the front of the jaw to location 2. The moment of force is now two instead of one, because the distance from the jaw joint is now longer (1×2 instead of 1×1). If the jaw remains in equilibrium, the bite force moment, for any tooth, must also be larger. Therefore, at equivalent distances from the jaw joint, all the bite forces at the teeth are now higher. Figure 1.4 demonstrates this effect. Here, the lower curve is repeated from Figure 1.3, in which the muscle force is located one unit from the jaw joint at location 1. The upper curve indicates bite forces for a vector that is located two units from the joint, giving a moment of 2. The bite forces in the upper curve are higher at all equivalent locations along the jaw.

Postulating vectors that are progressively farther from the joint produce progressively larger moments and therefore higher bite forces. Those vectors closer to the joint produce lower bite forces. This implies that if high bite forces are important to an animal, the muscle moment should be as large as possible; that is, a postulated vector should be located as far forward as possible, as far from the joint as possible, where the mechanical advantage is better.

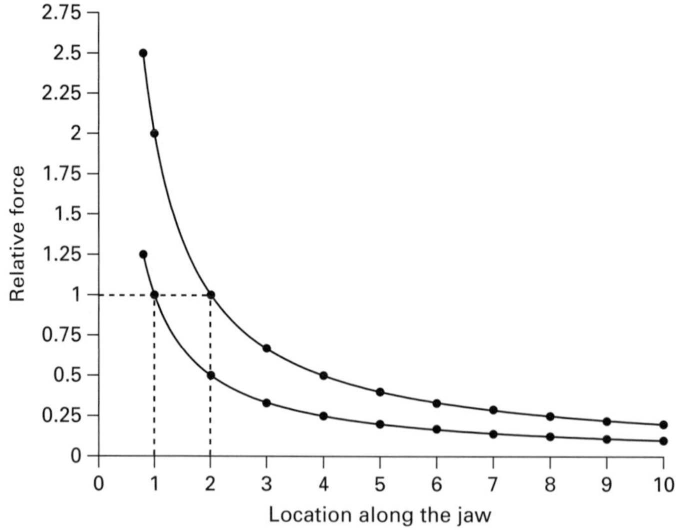

Figure 1.4. The muscle force in the lower plot is positioned at location 1. The muscle force in the upper plot is positioned at location 2.

But as noted, typically most of the muscles are not very far forward along the jaw in mammals. In fact, as will be discussed in detail at a later point, it is the case that the muscle force vector is approximately a third of the way along the jaw, giving a moment of near 3 in virtually all mammals. Therefore, many potentially high bite forces are for some reason not available to mammals. Apparently, other advantages accrue to these animals when the jaw muscles are located farther to the rear in a seemingly less beneficial location. This arrangement further suggests that other features of the jaw are more significant than bite forces that are as high as possible. In this regard, gape limitation and muscle interference with the mouth opening, as described above, hardly seem important enough.

Moreover, one expects the tooth row to be found in a mechanically advantageous location. Given the shape of the graphs, the rear of the tooth row should be as close to the jaw joint as possible, so that the bite forces at the posterior teeth, as well as at all the other teeth, will be as high as possible. Yet, exactly the opposite is observed in real mammals. There is virtually always a large toothless gap just in front of the jaw joint. These observations call attention to the fact that the analyses in Figures 1.3 and 1.4 are only part of the story; the reaction forces at the jaw joint have not yet been considered in enough detail.

1.5 Joint forces

In Figures 1.3 and 1.4 the graphs indicate the amount of *bite* force at each point along the tooth row. Another curve, for the *joint* force, can be added to plots like this. Figure 1.5 includes both a curve for bite force similar to those in Figures 1.3 and 1.4

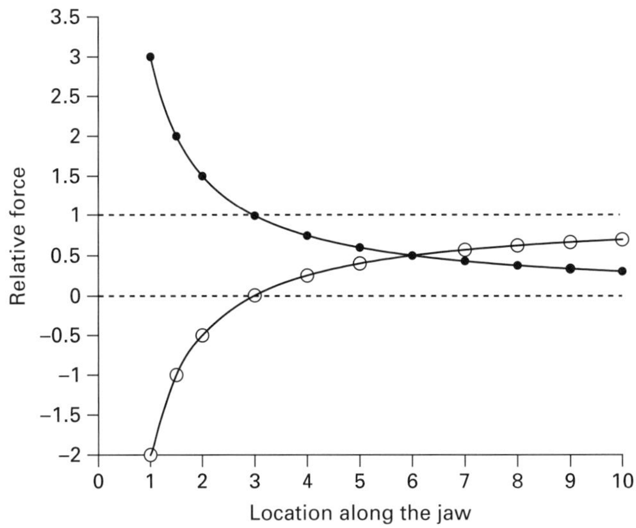

Figure 1.5. In this example the muscle force is positioned at location 3. The curve that is concave upward (filled circles) indicates relative bite force. The curve that is concave downward (open circles) indicates relative joint force.

(filled circles) and a curve for joint force (open circles). The muscle force again has a magnitude of one unit. The muscle force in this idealized example is imagined to be located three units from the jaw joint at location 3, giving a muscle moment of 3, and the joint is again located at location 0 on the horizontal axis. Recall that when biting at any point along the tooth row, the *sum* of the bite force and the reaction force at the joint is equal to the muscle force. Thus, adding the values on each curve at any given point along the jaw results in a total of one unit. This is the case because both the bite and joint forces are simply reactions to the muscle force.

It follows that only one of the plots in Figure 1.5 is really needed. For example, if only the bite force curve is present, the distance from the horizontal axis (at 0) *up* to the curve indicates the magnitude of the bite force at the chosen point. At the same time, the distance from the dotted line, located at *one* unit of force, *down* to the curve indicates the value of the joint force at the same point. Because of these relationships, either curve can be rotated 180° around a horizontal axis and then superimposed on the other.

The *bite* force curve rises from a low value (0.3) at the anterior end of the jaw at location 10 to very high values at the rear near the jaw joint. The *joint* force curve has to be examined more carefully. The *joint* force curve has its highest value of 0.7 at location 10 and falls to 0 at location 3. The value of 0 reflects the fact that the bite point is located directly under the hypothesized muscle force at location 3 and so the bite point resists all the force and the joint resists no force. Posterior to location 3 the *joint* force becomes negative.

When biting at, say, the anterior end of the jaw, the muscle force is pulling the jaw up while the two reaction forces, one at the bite point and one at the jaw joint, are resisting the upward movement by pushing the jaw down. Upper and lower teeth at the bite point, and the two halves of the jaw joint, are being forced together. These jaw structures are therefore being compressed. In an equilibrium condition, when the closing jaw meets resistance and stops moving, the muscle force is equal to the sum of the bite force plus the joint force. This is true for all bite locations, but anterior to location 3 the forces at the joint are always positive on the graph. The skull pushes down to keep the lower half of the jaw joint from moving up.

Posterior to location 3, closer to the back of the jaw, the joint force is negative on the graph and becomes more so at more posterior bite locations. These joint forces are not getting lower and lower, they are getting higher and higher. When the joint force is positive on the graph, the jaw joint surfaces are pushed together and a *compressive* force is resisted by the upper joint surface on the skull (the skull is pushing down). When the joint force is negative on the graph, the joint surfaces are pulled apart and a *tensile* force must be resisted. Now the skull has to pull up on the lower jaw, via the joint ligaments, to keep the joint from being dislocated. In this case the jaws are effectively functioning like a nutcracker where the joint capsule and ligaments take the place of the strong rivet that is required to resist strong tensile forces.

These observations from Figure 1.5 can be presented from a more mechanical point of view. The upper panel in Figure 1.6 indicates one hypothetical relationship between: (1) the jaw joint (circle at the left); (2) the muscle vector (vertical arrow); and (3) the tooth row (long, narrow rectangle). This panel in the figure shows a hypothesized muscle vector intersecting the tooth row so that some of the posterior teeth are located behind the muscle vector. The filled circle that is located just behind the muscle vector represents a food item. As the muscles pull on the jaw, rotating it around the jaw joint, the front of the jaw rotates up. During this movement, the food eventually comes into contact with both an upper and lower tooth. At that time, the food in effect becomes the fulcrum, or the point around which the jaw rotates; the front of the jaw continues to move up, but the back end now moves down. The horizontal arrow in Figure 1.6 (upper panel) indicates where this movement takes place at the jaw joint. In the case represented in this diagram, strong bite forces are not effectively applied to the food, because as the front of the jaw is pulled up the joint is simply pulled apart as the jaw rotates around the food item, and very small forces, if any, are applied to the food.

However, if a sufficient upward reaction force is applied at the jaw joint by, say, the joint ligaments, the joint is then not disarticulated, or pulled apart. Then as the jaw rotates around the jaw joint, strong bite forces are applied to the food. This cannot happen unless the reaction force at the jaw joint is in an *upward* direction so that the lower half of the joint is pulled up rather than pushed down. The capsule and ligaments of the jaw joint, perhaps some muscle fibers, and rarely a protuberance of bone that extends underneath the condyle to some extent, are the only anatomical structures that are potentially available to provide this force in mammals.

The multiplying effect of a lever system can produce very high bite forces. This occurs when a jaw lever functions like a nutcracker, where the magnitude of

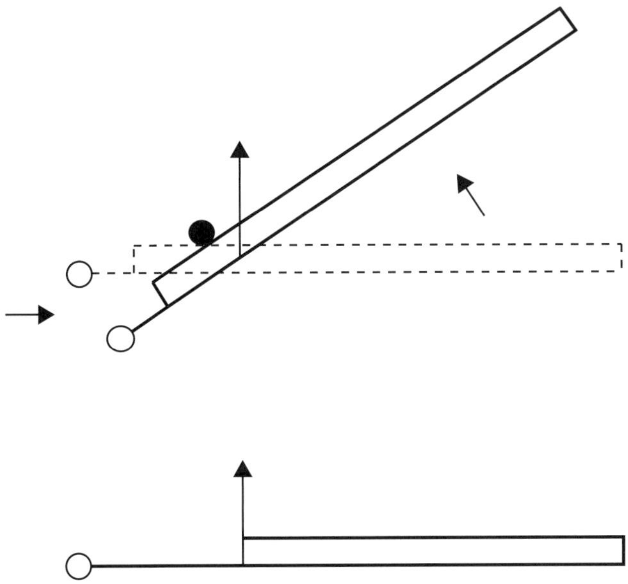

Figure 1.6. Hypothetical relationships between the jaw joint (circle), the muscle force (vertical arrow), and the lower tooth row (rectangle) are indicated. Above, the jaw has rotated around a resistant food item (filled circle) dislocating the jaw joint (indicated by the horizontal arrow). Below, the tooth row is anterior to the muscle force.

the (output) bite force is higher than the magnitude of the (input) muscle force. However, when this happens, the capsule and ligaments of the jaw joint must resist very high tensile forces. Anterior to the muscle vector, the magnitudes of both the bite and joint forces are modest, in that they are almost always lower than the muscle force. Posterior to the muscle vector the multiplying effect of the jaw lever is active, and these magnitudes are larger than the muscle force and become extremely high close to the jaw joint. Thus, at the region of the graph at the back of the jaw, where the joint force has a negative sign, the joint force is directed down and away from the skull and so the joint surfaces are being pulled apart (Fig. 1.5). Thus, when biting anterior to location 3 the jaw joint is resisting relatively low *compressive* forces, whereas posteriorly, behind location 3, it is resisting relatively high *tensile* forces.

At bite points behind the vector, the very high tensile forces pulling the joint apart are only slightly lower than the high compressive forces at the bite point. The surfaces of the jaw joint are well designed to resist high compressive forces, as joints are compression-resisting, not tension-resisting, devices. That is one major feature of most freely movable, or synovial, joints. Usually the only anatomical structures available to resist high tensile forces are the joint capsule and its ligaments. However, these ligaments can resist only modest tensile forces. Although very high bite forces at the rear of the jaw might be advantageous during chewing, at the same time, their presence requires that the joint ligaments resist very high tensile forces. If biting takes place at even a short distance behind the muscle vector, these ligaments maybe disrupted. This implies that high tensile forces at the jaw joint will not be present in a real animal.

A muscle force located far forward, presumably because of anteriorly attached muscles, might interfere with the mouth opening as well as reduce jaw gape. However, high tensile forces at the joint seem to be the more significant factor. The best evidence for the lack of tensile forces at the jaw joint is that, generally, there are no anatomical structures available to resist this tension. Typically, oversized joint ligaments, a muscle, or a strong bony process that holds the joint together are absent.

Consider the *bite* point that is located at location 1 along the jaw (Fig. 1.5). The moment of *muscle* force in this case is 3 (the one unit of muscle force is located three units from the jaw joint ($1 \times 3 = 3$)). The moment of *bite force* is also equal to 3 because the jaw is in equilibrium. The *bite* force moment is equal to the magnitude of the bite force multiplied by 1 (the distance from the fulcrum). The bite force is therefore equal to three units of force. The upward muscle force of one unit produces a bite force of three units. Given an equilibrium condition, an upward *joint* force of two units is required in order to balance the upward and downward forces. In a lever like this, the tensile force at the jaw joint is equal to the bite force minus the muscle force. The joint forces associated with bite points directly behind the muscle force are modest, but they become increasingly high for bite points just a short distance back toward the joint. As the bite force gets higher and higher, the joint force does lag behind, but only by an amount equal to the muscle force (one unit). Very high tensile forces on the joint ligaments accompany very high bite forces at the rear of the jaw. This condition worsens as the bite point gets closer and closer to the jaw joint.

No *overly large* tensile forces at the joint are present if biting takes place only at locations *anterior* to the presumed muscle vector (cf. Figs 1.1 and 1.5); that is, if all the teeth are located in front of the vector. In that event, all the joint forces are modest compressive forces and smaller than, or equal to, the muscle force. In the lower panel of Figure 1.6 all the teeth lie in front of the muscle vector. This means that no matter where food is being bitten along this tooth row, the muscle force, when it is in this location, always pulls the upper and lower tooth rows, as well as the two halves of the jaw joint, together. The vector therefore, is always located along the jaw *between* the jaw joint and the last molar. The significance of this location of the vector, always between the joint and the bite point, is that excessive tensile forces on the jaw joint are completely avoided (cf. Figs 1.5 and 1.6). Therefore, as the joint capsule and ligaments generally cannot provide the required high-magnitude forces, and alternative structures are typically absent, all the teeth are expected to be located in *front* of the muscle force. The presence of the few cases (e.g. badger) where a process of bone at the back of the joint on the skull partially extends under the condyle to apparently resist some forces that could pull the joint apart does not invalidate the general observation. The possibility exists that this animal's burrowing habit, rather than the jaw mechanism *per se*, requires this atypical structure.

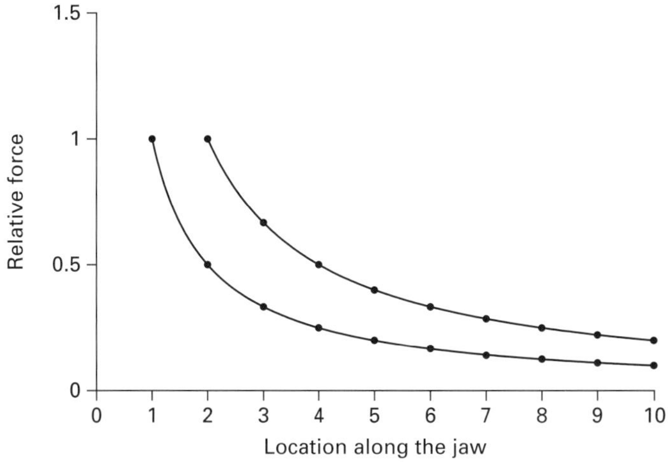

Figure 1.7. The muscle force in the lower plot is positioned at location 1. There are no bite forces behind location 1 because there are no teeth. The muscle force in the upper plot is positioned at location 2. There are no bite forces behind location 2 because there are no teeth.

The natural design of joints provides for compression resistance. Joints are typically not strongly pulled apart during their natural mode of operation. Moreover, a joint is more likely to be injured when disarticulating forces act on this structure. In so important a joint as the jaw, we might expect that the natural design of the masticatory mechanism would virtually preclude joint dislocation because of the increased possibility of injury. An animal in the wild that cannot chew properly is in very serious difficulty.

When all the teeth are in front of the muscle resultant force, there are no downward joint forces that tend to dislocate the jaw, and the skull can easily resist the modest upward, compressive forces at this joint. However, in this situation, there can be no multiplying effect by the jaw lever system, and bite forces are *less* than the muscle force in almost all cases.

If no teeth are located behind the muscle vector, and significant tensile forces in the joint ligaments are avoided, then the plots of bite force along the jaw (cf. Fig. 1.4) have to be modified. There can be no bite forces if there are no teeth, and so the plots are truncated at the rear, as in Figure 1.7. Here the lower plot goes no farther back than location 1 and the upper plot ends at location 2 because these are the locations of the muscle forces in these cases.

Accepting that the joint capsule and ligaments cannot cope with excessively large tensile forces and muscles or bony processes that could theoretically do this job are typically absent, the potential presence of such forces seems to be a very serious difficulty. Yet these tensile forces can be precluded in a remarkably simple way. Positioning all the teeth anterior to the resultant muscle force completely eliminates the problem. When there are no teeth behind the muscle vector, all the forces at the joint are compressive, although this force could theoretically approach 0. There are rarely any tensile forces at the jaw joint (Fig. 1.5). The unavoidable *disadvantage* of this arrangement is that it eliminates some very large bite forces because the multiplying effect of the jaw lever system is not utilized. High bite forces at some of the posterior teeth might well be advantageous, but the cost of these higher forces is the high tensile forces in the joint ligaments. This is a cost that is most probably too high.

Again, the best evidence that teeth are in front of, rather than behind, the muscle force is the observation that generally there are no anatomical structures capable of resisting very large tensile forces at the jaw joint. If teeth were located behind the presumed muscle vector, some means of providing an upward force at the jaw joint would be required. The joint ligaments are rarely large, powerful structures; only occasionally do large post-glenoid processes wrap around the condyle to prevent dislocation (moreover, such a structure is very weak in a mechanical sense), and large muscles that could theoretically hold the joint together are absent. (Some of the posterior muscle fibers of the temporalis and masseter are close enough to the jaw joint that they potentially could function in this way to some extent.)

1.6 The length of the tooth row

The foregoing relationships necessitate a very serious restriction on the jaw mechanism. If the jaw has a fixed length and all the teeth are anterior to the imagined vector, then the total length of the tooth row must be shorter if the muscle vector is farther forward, and longer if the vector is farther to the rear. That is, the length of the tooth row varies directly with the location of the muscle force vector. A muscle force located near the anterior end of the jaw necessitates a relatively short tooth row, whereas a posterior force allows for a relatively long row.

Recall two of the observations that were mentioned in the Introduction: first, there is almost always a region lacking teeth just in front of the jaw joint; and second, the jaw muscles are located relatively far back along the jaw. If the teeth are always located or positioned in front of the presumed muscle vector, these observations make perfect sense.

First, there is a toothless region in front of the joint because contraction of the muscles will close the jaw only if the resultant muscle force is located at some point anterior to the jaw joint. Further, a muscle force a short distance in front of the jaw joint will not have sufficient mechanical advantage. Therefore, the muscle force is expected to be far enough in front of the joint so that reasonable bite forces will be produced (cf. Fig. 1.4). If the teeth must be in front of the muscle force then there must necessarily be a region without teeth directly in front of the joint. Regardless of the actual location of the muscle resultant, it must still be some distance in front of the joint, and the teeth will then be slightly forward of that point.

Second, most mammals have a fairly long tooth row made up of four different kinds of teeth: incisors, canines, premolars, and molars. A tooth row of reasonable length is then possible only if the muscle force is located far enough behind the anterior end of the jaw. There would be room for only a very short tooth row in front of a muscle force located far forward if it is accepted that all the teeth are located in front of this force. Thus, a muscle force that is relatively far back along the jaw is necessary for a reasonably long tooth row, and a muscle force located far forward along the jaw is unlikely, because that leaves too little space for teeth. Accepting these points, the relatively posterior attachment of muscles in mammalian jaws and the resulting posterior location of the resultant muscle force is *required* if a reasonably long tooth row is present. The above considerations suggest that the observed muscle and tooth locations in most mammalian jaws are logical necessities accepting the suggested constraints on the system.

The analysis so far does not predict a particular location for the resultant force of the jaw muscles. Rather, it predicts that there will be two regions where the muscle force will *not* be found: at the back of the jaw just in front of the jaw joint and at the very front of the jaw. Thus, by default, the analysis predicts that the resultant force will be located at an as-yet-undetermined distance from the jaw joint in the middle of the jaw. In a very general way, this result conforms to the actual observations on mammals that were described at the outset. The next step in the analysis is to refine this result so as to predict a specific location for the muscle force.

1.7 The location of the muscle vector

Bite forces are higher when the muscle force is farther forward because the mechanical advantage is better. However, the tooth row is necessarily shorter in these cases. A longer tooth row is paired with lower bite forces because the muscle force is farther to the rear, where the leverage is not as good. These considerations focus our attention on two variables: *length* of the tooth row and *magnitude* of the bite forces. As magnitude increases, length decreases and *vice versa*; both cannot be maximized at the same time. Neither high bite forces along a tooth row that is too short nor a long tooth row with forces that are too low seems optimal. Moreover, the inverse relationship between length and magnitude implies a location for the muscle force that represents the best numerical compromise between these two variables.

Another difference between a long and a short tooth row is the number of potential points along the row where a bite force can be applied. A long tooth row can apply many forces, but they are all relatively low. A short row can apply fewer forces, but they are all relatively high. These considerations imply that both the length of the tooth row and the magnitude of the bite forces can be *combined* in a meaningful way by measuring the *sum* of the bite forces along the entire tooth row. (This is not to imply that all the teeth are used at exactly the same time. Rather, only that over a day, a week, or even a longer period, all the teeth are likely to function.) This sum of the bite forces along a row turns out to be equivalent to the area under the bite force curve (cf. Fig. 1.7). The reason for this can be seen by examining an elementary analysis that uses the sum of a series of rectangles to estimate the area under a curve.

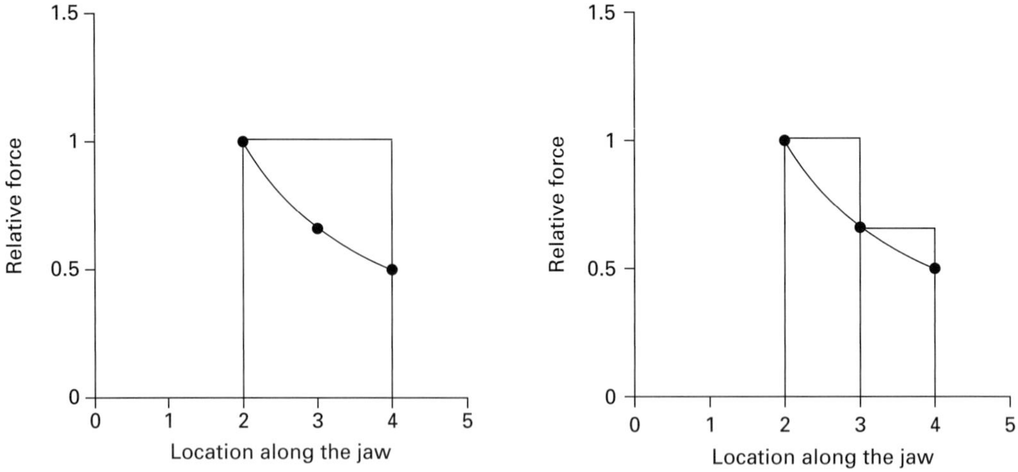

Figure 1.8. Left, the area under the curve estimated with one rectangle. Right, the area under the curve estimated with two rectangles.

How the sum of the forces along such a bite force curve can be determined is perhaps not intuitively obvious. Consider the hypothetical plot of bite force (Fig. 1.8, left panel). Enclosing this curve with a rectangle allows a very rough estimate of the area under the curve to be easily made. The length and width are simply multiplied together, which cannot be done when a curved line is on one side of the figure. The bottom of the rectangle is the x-axis and the two sides are verticals passing through each end of the curve. To complete the top of the rectangle, a horizontal line is drawn from the highest point on the curve to the vertical at the other end. The area of this rectangle has an area that is much larger than the actual area under this curve. The excess area lies within the rectangle but above the curve.

Drawing two rectangles instead of one and summing both their areas gives a better estimate of the area under the curve, even though this estimate is still poor (Fig. 1.8, right panel). The sum of the excess areas in this case is smaller than in the case where only one rectangle was used (Fig. 1.8, left panel). Larger and larger numbers of rectangles can be used to estimate of the area under the curve. As the number of rectangles is increased, an increasingly better estimate of the area is produced because the excess area, within the rectangle but above the curve, becomes smaller and smaller.

At the same time, the width of each rectangle necessarily decreases as their number increases. If enough rectangles are considered, their widths become so short that each rectangle essentially becomes a vertical line. Moreover, one of these lines (which is really a very narrow rectangle) is then simply the distance from the x-axis up to the curve. This distance indicates the magnitude of the bite force at that point on the curve. All of these magnitudes, at all the points on the curve, taken together represent the sum of all the forces along the curve. Adding up all these magnitudes is the same thing as adding up all the narrow rectangles. As the sum of the rectangles is the area under the curve, this area is also the sum of the magnitudes.

The above is nothing more than a procedure typically considered in elementary calculus. The area under the curve, or sum of the forces, can be measured graphically in a number of different ways, but it is more satisfactory to calculate the area by simply integrating the equation of the curve (see Box 2.1 in Chapter 2 for the calculations used to determine the areas under the curves).

Calculating such a sum of forces allows a comparison among different plots of bite force, where each represents a different location for the idealized muscle vector. This procedure combines both bite force and tooth row length (Fig. 1.7). This approach does not imply that an animal uses every conceivable bite point during each chewing cycle, which is clearly not the case. However, over a period of time, and considering all biting activities, all bite points will be used eventually, even though every tooth or biting region may not be used exactly the same number of times. Thus, a consideration of the *sum* of all forces seems warranted. The following analysis will continue to utilize a ten-unit-long jaw that extends from the jaw joint to the anterior incisors with a muscle force of one unit. For each different location of the muscle force (cf. Fig. 1.7), the magnitudes of the bite forces *at all bite points along the tooth row* can be summed by integrating the equation of the line from the bite force location out to the anterior end of the jaw. Recall that there are no bite forces behind the muscle force, because there are no teeth in those locations.

A series of bite-force graphs, each with a muscle vector located at a different location along the jaw, can be plotted (cf. Fig. 1.7). In each case a muscle force with a magnitude of 1 unit of force is assumed. The sum of the forces is naturally different in each case because each plot has a different length and the forces along each graph are different for equivalent locations along the jaw. The number of forces is large when the muscle force is located close to the jaw joint because the tooth row is so long. At the same time the leverage is poor, so the bite forces all along the row are relatively low. Not surprisingly, the sum of bite forces turns out to be relatively low as well. The sum is also low when the muscle force is located *far* forward along the jaw. Initially this might appear surprising, given that moving the muscle force toward the front of the jaw increases the distance from the jaw joint, thus improving the leverage of the muscle force, and therefore increasing the bite forces. The bite forces are indeed higher when the muscle force is *far* forward. However, because all the teeth must be in front of the muscle force, the tooth row is very short. Summing a small number of high forces results in a sum that is relatively low.

The largest sum of forces is found to be associated with an idealized muscle force that is located slightly more than one-third of jaw length from the jaw joint. This intermediate location is close to the actual estimated location seen in those mammals that have been examined. A muscle vector in this location maximizes the sum of the output bite forces at the teeth all along the tooth row. This is true regardless of the absolute magnitude of the muscle force that is put into the system. Accepting that bite force and tooth row length are both important attributes of the jaws, this result is of great interest because it provides a reason for the actual location of the muscle force; the sum of the bite forces is maximized. Moreover, estimates indicate that a similar location of the muscle force is also seen in the earliest mammalian fossils. Thus, this relationship between the location of the muscle force and tooth row length was present at the origin of this group of animals and perhaps before. The chain of analyses described above suggests that this location is probably inevitable, as it maximizes the sum of the output bite forces consistent with the constraints on the mammalian masticatory apparatus.

Maximizing the sum of the bite forces requires that teeth fill the entire distance from the muscle force out to the anterior end of the jaw. If teeth are missing, there can be no bite forces in those locations. The sum of bite forces in such a case will naturally be somewhat lower than in a situation in which the entire distance is filled with teeth.

To add clarity to the discussion above, three specific cases can be examined from a slightly different point of view. In each of these cases a muscle force with a magnitude of one unit is assumed. The muscle forces are located at location 1, location 3, and location 9 (cf. Fig. 1.7). A muscle force that is located at location 9 has better leverage than if it were located closer to the joint, because the muscle moment arm is so long (Fig. 1.9a). The tooth row is necessarily rather short because all the teeth must be in front of the muscle force, and thus the bite force plot extends from location 9 to location 10 on the horizontal scale. The area *under* the curve represents the total of all the bite forces along this anteriorly positioned, but very short, tooth row. The bite forces in this case are all reasonably close to one unit of force. Recall that the bite force plus the joint force equals the muscle force. If the joint force curve were plotted, it would be the same shape as the bite force curve but turned upside down and positioned near the x-axis. Therefore, the small area *above* the curve in the upper right corner of Figure 1.9a indicates the sum of the joint forces. All these joint forces are rather low, each being much less than one unit of force. The bite reaction forces are large and the joint reaction forces are small. This would be a very efficient arrangement if tooth row length were unimportant.

A muscle force located at location 1 has much poorer leverage because of its short moment arm (Fig. 1.9b). This is clearly inefficient, because most of the muscle force is dissipated at the jaw joint rather than being used at the teeth. Note that the area above the curve, representing joint force, is much larger than the area below the curve, representing bite force. Almost all the bite forces are lower than the corresponding joint forces. The tooth row, however, is very long.

The final example locates the muscle force at location 3 (Fig. 1.9c). The nearly equal size of the areas above and below the curve indicates that approximately half the muscle force is wasted, whereas half is used constructively as bite force. At most locations along the curve where a comparison can be made, the magnitude of the bite force has an intermediate value compared to the two previous examples. Both the efficiency as well as the length of the tooth row has an intermediate value compared with the two previous cases.

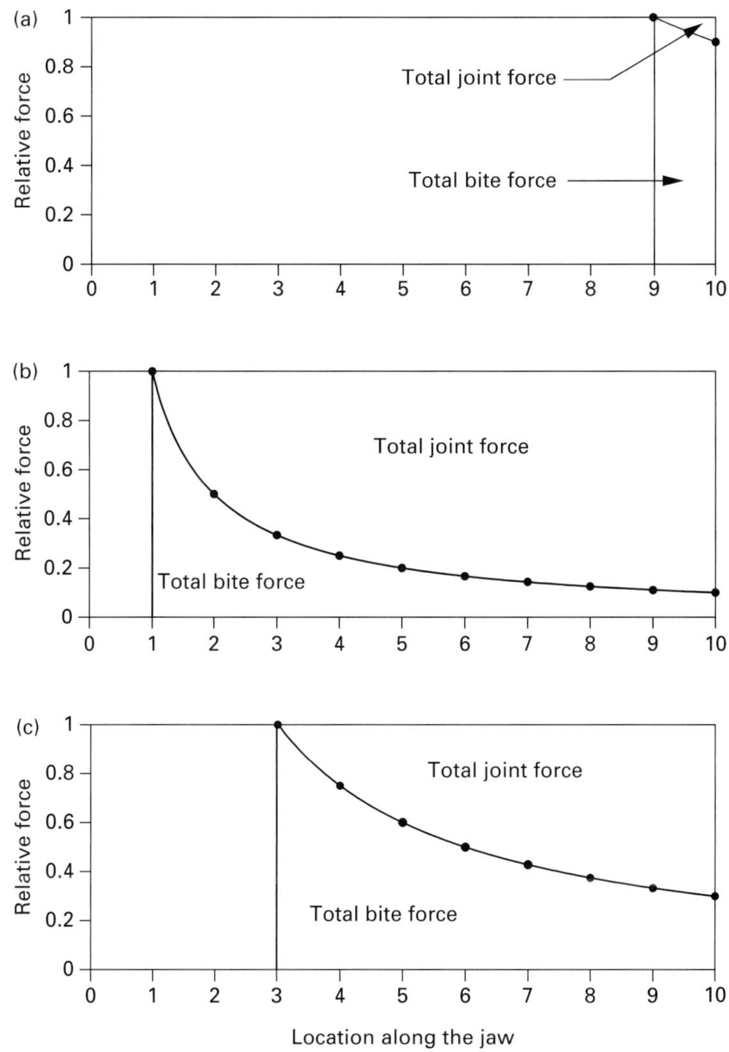

Figure 1.9. Plots of relative bite and joint force against location along a ten-unit-long jaw. The muscle force of one unit is located at location 9 (a), at location 1 (b), and at location 3 (c). Areas under the curves represent total bite force. Areas above the curves represent total joint force.

The important difference between these three cases is neither the efficiency (sum of the bite forces divided by the sum of the muscle forces) nor the length of the tooth row *per se*. Rather, the issue is the magnitude of the sum of the bite forces. This sum in any given case is related to both efficiency and tooth row length. As indicated, the first case (Fig. 1.9a), with the very short tooth row, is far more efficient than the other two. However, the absolute sum of the bite forces (0.95) is very small even though it is a very large fraction of the total input muscle force (1.0). Recall that the area above the curve is the sum of the joint forces, and thus the areas above and below the curve, taken together, equal the sum of the input muscle forces.

The least efficient example is clearly better than this, because the sum of the bite forces (2.3) is twice as large (Fig. 1.9b). The efficiency is low because this amount is only one-quarter of the total input muscle forces of 9.0 and the sum of the joint forces is 6.7. In this case the tooth row is very long.

The last example is not particularly efficient, as the sum of the bite forces (3.6) is only half of the total muscle input forces (7.0) and the sum of the joint forces is 3.4. However, this is the most advantageous condition of the three because the bite force sum is the largest. It approaches twice the size of the sum in the previous case. In addition, the tooth row is still reasonably long (Fig. 1.9c).

Comparing these three graphs perhaps gives a better visual impression of the reason the last case is better. A large fraction of a small total (Fig. 1.9a) is not as good as a small fraction of a large total (Fig. 1.9b) and neither is as good as an entire half of a total of intermediate size (Fig. 1.9c).

Recall that the jaw in these analyses has been considered to be two dimensional. The jaw mechanism, of course, is actually a three-dimensional structure. The best location for the muscle resultant in the two-dimensional case is greater than one-third of the jaw length from the jaw joint. The best location in a three-dimensional jaw turns out to be slightly different and is somewhat less than one-third of the jaw length from the jaw joint. The actual location in the three-dimensional case is 30% (0.30) of the way along the entire ten-unit-long jaw (discussed in detail in Box 2.1 in Chapter 2).

Finally, the major importance of the analyses up to this point is that the results conform to what is apparently seen in a large number of mammals. This suggests that the analyses have produced results that may provide explanations for why the masticatory apparatus has evolved the way it has. Although it is true that the location of the vector has not been measured yet in any mammal, good estimates from dissections and weighing of the jaw muscles of a number of representative animals all closely approach the results of the above analyses.

The above result is crucial because it indicates that the lever system in mammals is fixed at its most efficient configuration. The lever system is, in a sense, maximized. Therefore, bite force is determined by the amount of muscle tissue, as the lever system is not expected to change.

1.8 Summary

The jaw mechanism to this point has been analyzed heuristically as an extremely simple lever system. Bite forces and joint forces contribute to the analyses in different ways. A study of *joint* forces indicates where the teeth can be located along the jaw. Only after that analysis of joint forces was completed could the *bite* forces be examined profitably. The analysis of bite forces suggested the appropriate location along the jaw of the idealized resultant vector of jaw muscle force.

Although intuitively satisfying, considering bite forces alone at the outset gives the wrong answers. A study of bite forces predicts that jaw muscles should be located toward the front of the jaw to give the best leverage. Posterior teeth are predicted to be as close to the joint as possible in order to take advantage of the high potential bite forces in this region. Neither of these predictions is evident in real animals. Progress required considering joint forces first, at the very beginning of the analysis, before *returning* to a study of the bite forces.

Studying all possible bite locations reveals that *joint forces* can be compressive, tensile, or zero. The joint forces are compressive when biting takes place anterior to the resultant muscle force vector. In this case the joint surfaces are pushed toward each other. The joint forces are tensile when biting takes place posterior to the muscle vector. In this case the joint surfaces are pulled apart. The joint force is zero at the point where the biting tooth is located in the same place as the muscle force vector; compression turns to tension behind this point. When the joint forces are compressive they are modest, in that they are equal to, or lower than, the muscle force that is put into the system. When they are tensile they can be very large and tend to pull apart, or disarticulate, the joint and damage the joint ligaments. Finally, anatomical structures that are able to resist dislocation of the joint are not evident in most cases.

Biting that is restricted to locations where the joint forces are compressive avoids damaging the joint capsule and its ligaments. That is, all the teeth are located in front of the muscle resultant. Only at this point in the analysis can the magnitude of the bite forces be profitably studied. Because the length of the tooth row varies with the location of the vector, a vector located far to the rear allows for a very long tooth row. Bite forces are very low because of the poor leverage. A vector located far to the front of the jaw, with good leverage, produces high bite forces, but the tooth row is far too short (recall that all the teeth are in front of the vector). Thus, the vector must be somewhere in the middle of the jaw. One way of differentiating between different tooth row lengths is to consider the sums of the bite forces for different locations of the muscle vector. The location of the vector where calculations indicate that the sum of the bite forces, all along the entire jaw, is maximized is somewhat more than one-third of jaw length from the jaw joint.

[This result is for a *two-dimensional* jaw. In a more appropriate *three-dimensional* model, the bite force is maximized when the vector is located somewhat less than one-third of jaw length from the jaw joint. This case will be discussed in Chapter 2. In those cases where the location of a resultant vector has been estimated from dissections it has been found to be very close to this location (30% of jaw length from the joint).]

Given a fixed and most efficient lever system, the magnitude of the bite force is determined by the amount of muscle tissue available. Different lever systems are not equally efficient.

2 The jaw viewed as a three-dimensional lever

In a two-dimensional jaw mechanism, joint stability requires only that the resultant muscle force be located behind the most posterior tooth. A stable three-dimensional jaw mechanism requires that the resultant force of the jaw muscles be located within a triangle formed by the two jaw joints and the bite point. One consequence of this relationship is that the muscle force cannot be maximized at the rear of the tooth row and bite forces in this region are not as high as might be expected.

The anteroposterior location of the muscle force is 30% of jaw length from the jaw joint. The mediolateral location of the tooth row seems to be essentially the same in many mammals after taking the length and width of the jaw into account.

2.1 The three-dimensional jaw

In order to introduce a few elementary principles in a simple manner, the jaw up to this point has been discussed as though it were a two-dimensional object. In such a two-dimensional case, avoiding tensile forces at the jaw joint simply requires that teeth be in front of the muscle resultant force. However, the mammalian jaw mechanism is actually a three-dimensional structure, and this condition will now be taken into account.

Superficially, the mandible might seem to approach a two-dimensional object because one half of a lower jaw is often a relatively thin plate of bone. However, when the entire lower jaw, with its two halves, is considered, the mandible is clearly a three-dimensional structure. The skull is obviously three-dimensional. The jaw mechanism consists of two jaw joints, two sets of tooth rows, and two virtually identical groups of jaw-closing muscles. When an animal is chewing, especially during the power stroke of the chewing cycle, the muscles on both sides of the head are actively producing forces and both jaw joints are usually resisting reaction forces. Recall that these studies assume maximum bite forces when the teeth are at, or near, occlusion. Moreover, most mammals generally chew on only one side of the head at a time. During chewing there are no bite forces on the teeth on the non-chewing side. Because many jaw mechanisms function in this way, the lever system of a three-dimensional jaw turns out to be somewhat more complicated than it is in the simple two-dimensional case that has been discussed up to this point.

[The mandibular symphysis is fused in most primates (monkeys and apes) and in a small number of other cases, but most mammals have an open, or unfused, joint between the two halves of the lower jaw. The absence of a fused joint in most mammals suggests that fusion is not present to deal with typical masticatory needs such as transmitting muscle force from one side to the other. Certainly fusion makes for a strong and rigid joint between the two halves of the lower jaw. Further, a fused joint suggests that this condition is important to some mammals. However, a fused joint produces a lower jaw of single V- or U-shaped structure that is not strong in a mechanical sense. Thus, a good deal of bone tissue is required in order to buttress the region where the two halves of the lower jaw meet.

The above points suggest that whatever drives this fusion in some mammals is not important to most. During incisor biting at very short inter-occlusal distances, some of the jaw muscle force might be dissipated because if the joint is *not* fused, the upper and lower *incisors*, on the non-biting side, can come into contact. Muscle force used to force these teeth together is not doing useful work. Only approximately half of the total muscle force is then available to deal with food items that lie between the incisors on the other (chewing) side.]

When modeled as a two-dimensional structure, the lower jaw is represented as a bar with the jaw joint on one end, the teeth extending toward the middle from the other end, and with the muscle force vector located about one-third of jaw length from the jaw joint. As noted above, tensile forces that pull the joint apart are avoided completely if all the teeth are located anterior to the muscle force vector. When the entire lower jaw, with two tooth rows, is modeled as a three-dimensional structure, it can be represented as a triangular plate. The line connecting the two joints is the base, and the equal sides of the triangle pass along the tooth rows and meet at the symphysis at the anterior end of the jaw. Obviously, not all mammals have triangular jaws; a triangular plate is suggested as a very simple example that simplifies some calculations.

2.2 The triangle-of-support

When the lower jaw closes and approaches the upper jaw it rotates at the two jaw joints. The axis of rotation can be thought of as the line that connects these two joints. Other movements, especially sliding, also often take place depending upon the animal being examined and how the jaw is being used, but rotation is a major feature of jaw closing and will be emphasized here as a simplification, because the discussion is confined to jaws near occlusion.

On the side where biting is taking place, the jaw eventually stops moving when the upper and lower teeth come into contact or when they press against food that is positioned between them. Once this happens, the lower jaw is, in effect, in contact with the skull at only three places: the two jaw joints and the bite point. Within the joint cavity of each jaw joint, a disk usually separates the joint surfaces on the skull and lower jaw. The bite point is the region in which either the food item is caught between the teeth or in which the teeth actually come into contact. For ease of analysis, these three contacts will be considered to be *points* of contact, even though they are clearly not points, but rather small regions that have some areal extent. At any instant, these three points are the only places where the lower jaw comes into contact with the skull during strong biting.

Because there are three points of contact between the lower jaw and the skull, the lower jaw can be stable during biting. That is, the three points define a plane.

In the two-dimensional analysis, the three major jaw-closing muscles were resolved into a single resultant vector of muscle force that represented the combined effect of all three muscles taken together. In the three-dimensional model the three muscles on the other side must also be considered. These muscles, on the other side, can also be resolved into a single vector. As a means of simplifying the following analysis, the resultant vectors themselves, from each side, may be resolved into a single grand resultant that represents the sum of all the forces that the major jaw-closing muscles, on both sides of the head, are able to produce.

Figure 2.1 represents a plan view of the lower jaw from above as if the skull were transparent. The lower jaw itself is represented by the large isosceles triangle in light line where the two jaw joints (**J**) are at the ends of the two equal sides. A tooth at **B** in the left tooth row is chosen as a representative bite point. The triangle in heavy line (**JBJ**), which connects the two jaw joints and the tooth at the bite point, is the triangle-of-support . Two lateral muscle resultants are located at **M** on each side of the head. The grand resultant, which is derived from the two equal lateral resultants, is located on the midline at **R**. These resultants are seen end-on in this example and are therefore represented as points (filled circles). They would be seen as arrows in a lateral view.

A summary of the analysis to this point indicates that the single grand resultant force pulls the lower jaw up until it comes into contact with the skull at only three points, the two joints and the bite point. The skull provides downward forces at these three points. Note that the grand resultant force at **R** is bounded by, or lies inside, the triangle-of-support (triangle in heavy line).

An apt analogy for this situation, although upside down, is a three-legged stool. When sitting on such a stool, the force provided by a person's weight lies inside the triangle formed by the three legs. All three legs are pushed down toward the floor. The floor resists the downward force at each leg, so the stool is stable and does not tip over. If, however, the person shifts to one side, a position is eventually reached such that the person's weight no longer lies inside the triangle formed by the three legs. At this point the stool tips over. It tips over because the weight lies outside of the supporting structure of the stool that is formed by the three legs. As the stool tips over, one leg will move away from the floor. The stool will also rotate around an axis coincident with the line joining the ends of the two legs that are still in contact with the floor.

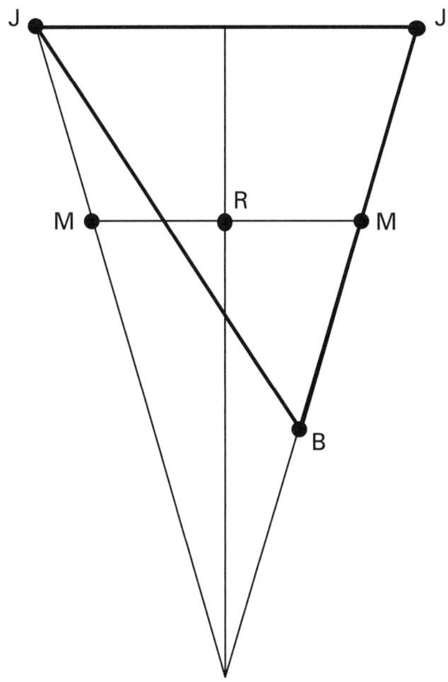

Figure 2.1. This figure is a plan view of the lower jaw from above. The jaw joints are at the top of the drawing and the anterior end is at the bottom. The jaw joints (J), muscle forces (M), the resultant muscle force (R), and a representative bite point (B) are indicated with filled circles.

The jaw is analogous to the stool in that the muscle force is equivalent to the weight on the seat, while the joint and bite forces are like the forces at the three legs. When the resultant muscle force is positioned outside the triangle-of-support the jaw also tends to rotate. As this rotation proceeds, one condyle tends to move down and away from the skull. This rotation will be prevented, up to a point, by the joint ligaments that are thus subjected to tensile forces. Dislocation of the joint is to be avoided, because stretching the ligaments dissipates some muscle force. But dislocation is to be avoided mainly because of the increased risk of damage to the capsule and ligaments of the jaw joint, should these forces become too large (see Chapter 1).

2.3 The distribution of joint and bite forces

The lower jaw is now thought of as a triangular plate (Fig. 2.1). A resultant muscle force tends to pull the plate out of the plane of the paper toward the reader. Three additional forces, one at each vertex of the triangle-of-support, resist the tendency of the plate to move up and are thus applied in the opposite direction. The upward force represents the grand resultant muscle force and the three downward forces represent the reaction forces at the two jaw joints and the bite point in this three-dimensional model. The jaw eventually reaches equilibrium so that it is not moving toward or away from the reader. The three downward forces are present in reaction to the single upward muscle force. The total magnitude of the three downward reaction forces is therefore equal to the grand resultant muscle force. The triangular plate remains horizontal, just as the lower jaw remains horizontal relative to a horizontal skull. The significance of the location of the resultant muscle force at **R**, inside the triangle-of-support, becomes evident when different *hypothetical* locations for the muscle force are considered.

Imagine a resultant **R** that is initially located at the center of the triangle-of-support so that it is equidistant from each jaw joint as well as from the bite point. It is intuitively obvious that the reaction forces at the joints and the bite point are all equal in this case. Suppose the resultant is moved toward one angle of the triangle-of-support, say, close to one of the jaw joints. In this case, the joint closer to the resultant force will have a high reaction force, whereas the other joint and the bite point will resist smaller forces because they are farther away from the resultant. A resultant force located directly over one angle of the triangle-of-support – say, over the bite point – is balanced by a single, and equal, downward reaction force at this bite point. The downward forces at the other two vertices, in this case, will therefore be zero. They must be zero because the downward forces, taken together, equal the upward resultant force.

As another example, consider a resultant force located at a point on, and near the middle of, one *edge* of the triangle-of-support. Imagine that one joint and the bite point define this edge. Compare this case with the analogy of the three-legged stool described above. The stool is about to tip over, the force at one leg is zero, and the two remaining legs resist the entire weight on the stool seat. In the jaw in this example, the bite point and one joint resist all of the muscle force and the remaining joint resists none of this force. If the resultant force were equidistant from the joint and the bite point, the forces at these two points would be equal.

As a final simple example, consider a resultant force located at a point near, but not on, one edge of the triangle-of-support. Three forces then balance the resultant force. There are now relatively higher forces at each end of this edge defined by, say, one joint and the bite point and a small force at the other joint.

Therefore, almost any combination of relative joint and bite forces is *theoretically* possible if it is assumed that the resultant force can be moved to any location within the triangle-of-support. This exercise demonstrates that the location of the resultant, within the triangle-of-support, determines the reaction forces at the two jaw joints as well as the reaction force at the bite point.

In real animals, the grand resultant is actually located at some point along line **MM**, because this line joins the two lateral resultants (Fig. 2.1). These two lateral resultants, one at each **M**, appear as points because they are viewed end-on. Each **M** lies about one-third of the way along the jaw, as previously described. The grand resultant can lie anywhere along the line connecting **M** and **M**. Exactly where on this line the resultant will lie depends upon the magnitudes of the two lateral resultants, which in turn depend on the activity of the muscles on each side of the head. If the lateral resultants are equal in magnitude, the grand resultant will be located at **R**, the midpoint of the line connecting the two lateral resultants. Again, the grand resultant is always equal to the sum of the two lateral resultants. If the magnitude of one lateral resultant is less than that of the other, the grand resultant is shifted toward the side where the larger lateral resultant is located. If the lateral resultant on one side is zero, the remaining lateral resultant is the grand resultant; in this case, the grand resultant obviously has the same magnitude as the single lateral resultant and is located in the same place.

Perhaps the simplest way to calculate the *location* of the grand resultant force is to note that the ratio of the *magnitudes* of the muscle forces is equal to the ratio of the *distances* from the grand resultant to each lateral resultant (**M**). This point will be discussed in more detail below.

[The lateral resultants can move anteriorly and posteriorly to some extent because of differential activity at the back and front of the muscle masses. Thus the grand resultant can also move anteroposteriorly to some extent. However, this anteroposterior movement of the grand resultant is much less than the lateral movement just described and so will be ignored in the following discussion.]

Two of the vertices of the triangle-of-support, the two jaw joints, are fixed in position, whereas the location of the bite point varies. Each different tooth, or small group of teeth, that serves as a bite point determines a different triangle-of-support. If biting takes place on a more anterior tooth, the triangle-of-support is longer than the supporting triangle (heavy line) illustrated in Figure 2.1. The distance from the resultant on line **MM** to this new, more anterior, bite point is longer. The distance from the jaw joints to the bite point is also longer. The bite force is lower in this case because a longer distance out to the bite point must be paired with a lower force to maintain equality with the input moment of muscle force, which has remained the same, as discussed in detail in Chapter 1. A tooth farther to the rear forms a shorter triangle-of-support and is closer to the resultant, and the bite force is higher. The joint and bite forces taken together equal the grand resultant in an equilibrium condition. If the bite force were low because of a more forward location of the bite point, one or the other, or both, of the joint forces would be higher to balance the muscle resultant. The opposite would be true if the bite force were higher. Different bite points define different triangles-of-support, and the magnitudes of the relative bite and joint forces change. They can be calculated if the position of the grand resultant, relative to the jaw joints and the bite point, is known (see below).

2.4 Tensile forces at a jaw joint in the three-dimensional case

Tensile forces at the single jaw joint in a two-dimensional jaw, as previously described, occur when biting takes place on a tooth that lies behind the muscle resultant force and the joint is pulled apart. If any teeth are located behind the muscle vector, a temporary fulcrum can form at one of these teeth. In such a case, high forces at the jaw joint will tend to dislocate the joint as the jaw closes and rotates about this temporary fulcrum. The presence of two jaw joints, even in the simplest three-dimensional

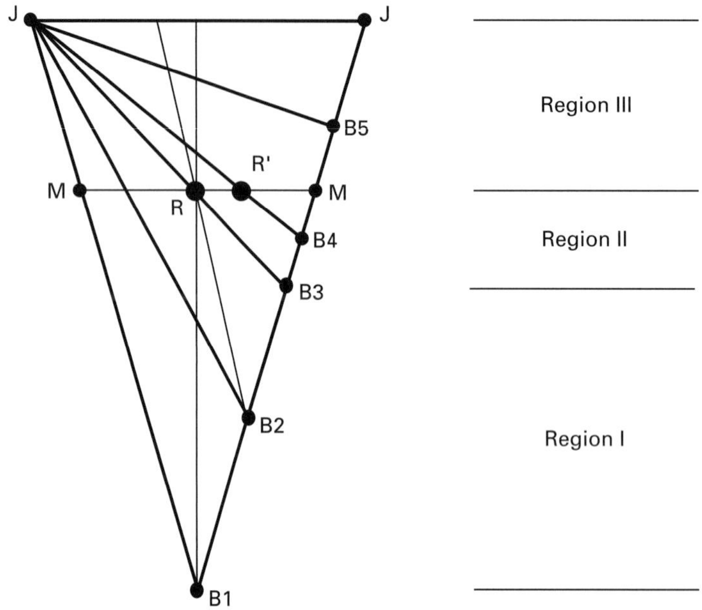

Figure 2.2. This figure is a plan view of the lower jaw from above. Five representative bite points, B1 through B5, are indicated with their respective triangles-of-support in heavy line. Regions I, II, and III are indicated.

case, makes for a more complicated situation. In the simplest three-dimensional case the issue is not where the vector is placed relative to the tooth row, but rather where the vector is positioned relative to the triangle-of-support.

A drawing similar to that in Figure 2.1 will make this clear. Point **B1** represents a bite point at the incisor region (Fig. 2.2). The triangle (**JB1J**) that represents the entire jaw also represents the triangle-of-support in this instance. The grand resultant muscle force is located at **R** on the midline of the jaw. The location of the grand resultant on the midline at **R** indicates that the lateral muscle resultants on each side of the head at **M** and **M** are equal. As in the previous examples, the resultant at **R** is equal to the sum of the two joint forces plus the bite force. In this case, the two joint reaction forces at **J** and **J** are equal because the jaw lever is the midline (see below). The reaction force at the bite point **B1** is somewhat less than the force at either joint because the bite point is farther away from **R**. Clearly **R** is inside the triangle-of-support and so the jaw is stable.

If biting takes place a little farther back along the tooth row at, say, **B2**, the triangle-of-support changes shape and is now **JB2J**. The resultant at **R** still lies inside this supporting triangle and the bite force increases somewhat, because the bite point is closer to **R** and to the joints. When biting at **B3** the triangle-of-support is **JB3J** and is again a different shape. The resultant at **R** is still inside the triangle-of-support, but just barely so this time, because it now lies on the line **JRB3**, which forms one side of the supporting triangle.

As each of these cases is considered in turn, the bite force is seen to increase because the bite point gets closer to **R** and to the joints. The joint forces also change because one side of the supporting triangle (from the right side joint to **B2** or **B3**) gets closer to the resultant muscle force (at **R**). The reaction forces at each *end* of these sides of the triangles, at the bite point and the joint on the non-biting (right) side, become larger. The force at the joint on the *non-biting* side therefore *increases* relative to the force at the biting side joint (left side). This is the case because **R** is closer to one side of the triangle-of-support. In all cases, the sum of the downward reaction forces at the jaw joints and the bite point resist the upward pull of all of the jaw muscles. If biting takes place behind **B3**, the jaw mechanism functions in a fundamentally different way. **B4** is such a bite point.

In this case the position of the triangle-of-support relative to the resultant is different. **B4** is far enough back along the tooth row so that the triangle-of-support (**JB4J**) is located entirely behind **R**. Thus **R** lies outside, rather than inside, the supporting triangle.

Recall again the analogy of the three-legged stool. If the person's weight lies inside the triangle formed by the three legs, all the legs are pushed down toward the floor. The floor resists each of these forces and the stool is stable and does not tip over. The stool tips over if the weight is shifted far enough to the side so that it no longer lies inside the triangle-of-support formed by the three legs. One leg moves away from the floor as the stool rotates around the axis formed by the line joining the two remaining legs that are still in contact with the floor. The jaw rotates around the line that connects the bite point at **B4** with the right jaw joint, which is on the opposite, or non-biting, side of the jaw (Fig. 2.2). The condyle of the joint on the left, the biting side in this case, tends to move down and away from the skull, tensing the joint ligaments.

The point of Figure 2.2 is to illustrate the location of those bite points that will cause the jaw joint on the biting side to be disarticulated and those bite positions that will not cause dislocation. Bite point **B4** is far enough back that **R** is well outside the triangle-of-support. The jaw is unstable and the jaw joint on the biting side will be dislocated. Bite point **B2** is far enough forward that **R** is well inside the triangle. In this stable situation, both joints as well as the bite point are pushed together and therefore resist compressive, not tensile, reaction forces. The biting side joint will not be dislocated. Bite point **B3** is just on the dividing line between the region of stability and the region of instability. The jaw joint on the biting side will not be dislocated when biting at **B3**, but technically, in this geometric model, this joint will resist a reaction force of zero. Therefore, the biting side jaw joint is just on the verge of being dislocated. This joint will be dislocated when biting takes place at any point behind **B3**, but not when biting takes place at any point in front of **B3**. There is nothing mysterious about bite point **B3**. It simply defines a triangle-of-support such that **R** is located, exactly, on one edge of the triangle-of-support. As the jaw is symmetrical, these arguments apply to either side.

This analysis is presented as a geometric example and as such is extremely simplified and idealized. Presumably in a real animal a safety factor is involved, and this particular idealized condition will only be approached and biting is expected to occur slightly forward of **B3** in almost all cases. For example, in some reptiles, such as turtles, the jaw mechanism is structured with a beak (rather than teeth) that extends no farther to the rear than point **B3**. Recall that in this geometrical example, as well as in many reptiles, muscle activity on both sides is assumed to be equal so that the grand resultant is positioned at **R**.

In the two-dimensional case that was mentioned earlier (see Chapter 1) and that was also overly idealized, joint dislocation was avoided by simply positioning the muscle force vector behind the teeth somewhere between the jaw joint and the last tooth. Potentially, the tooth row could extend all the way back to the resultant force itself; the teeth only had to be in front of the muscle force. In the somewhat more realistic, but still highly idealized, three-dimensional case just described, the tooth row cannot extend any farther back than point **B3**, which is some distance in *front* of the resultant muscle force. Therefore, the last tooth in the row *in this example* is a significant distance anterior to the resultant muscle force that is located at **R** on line **MM**, assuming stability is to be maintained. In Figure 2.2, this distance extends from **M** to **B3** along the biting side of the jaw. In a *triangular*, three-dimensional jaw, **B3** is located *approximately* halfway along the side of the jaw.

The preceding paragraphs discussed two very different regions along the jaw. One region extended from **B1** to **B3**. Dislocation of one of the jaw joints is almost impossible if the bite point lies in this region and the resultant is located at **R** (Fig. 2.2). The second jaw region extends from **B3** to **M**. The jaw is unstable when the bite point is in this part of the jaw, and the biting side joint tends to be pulled apart, endangering the joint ligaments. The regions from **B3** to **M** and from **M** to **J** will be discussed more fully below.

When biting at **B1** the entire jaw forms the triangle-of-support (**JB1J**) and the resultant muscle force is close to the middle of the supporting triangle (Fig. 2.2). If the resultant moves toward **M** on the biting side, the stability of the jaw is unaffected because the resultant will still lie inside the triangle-of-support. Even if the resultant were superimposed on **M** at the biting side, the jaw would still remain stable. Study of Figure 2.2 demonstrates that any bite point from **B1** to **B3** coupled with any location of the resultant from the midline position at **R** to **M** on the biting side produces a stable jaw. This is true because in virtually all of these cases the resultant lies inside the supporting triangle.

2.5 Different locations of the resultant force

A bite point in location **B4** in the example above caused instability only because the resultant muscle force was located on the midline at **R** (Fig. 2.2). However, as already mentioned, the resultant can be positioned at any point between **M** and **M**. Therefore, all that is required for jaw stability when biting at **B4** is a resultant located at least as far to the side as **R′** (Fig. 2.2). The resultant can actually be located at any point between **R′** and **M** (on the biting side). The resultant is then located either on the edge of the triangle-of-support, or inside it, and the jaw joint will not be dislocated, precluding the possibility of ligament damage. Again, to provide a safety factor in real animals, the resultant is probably located a little closer to **M** to avoid joint dislocation should the actual muscle forces vary slightly.

When mediolateral movement of the resultant is allowed, any bite point from **B3** to **M** is possible without causing jaw joint dislocation. The only requirement is that the resultant muscle force be located far enough toward the biting side of the jaw so that it is located inside the supporting triangle. Thus, the region from **B3** to **M** is *similar* to the more anterior region (**B1** to **B3**) in that stability of the jaw can be maintained, but is *different* in the sense that not all positions of the resultant are possible. There is an additional crucial difference.

When the resultant lies on the midline of the jaw, the lateral resultants at **M** are equal. For simplicity, the magnitudes of the lateral resultants are initially assumed to be at a maximum value; all the jaw muscles are exerting their maximum force. As will be described more fully below, if the resultant is located closer to the biting side of the jaw, to avoid dislocation of the biting side joint, the muscle force on the non-biting side must *necessarily* be lower than that on the biting side. The total grand resultant force is then also *lower*. Although **R** can be located at any point between **M** and **M**, the muscle force is highest at the midline, lowest at either **M**, and of intermediate magnitude elsewhere. When biting at **B3** and the resultant is on the midline, the largest amount of muscle force is potentially available. When biting at any bite point behind **B3**, say **B4**, the resultant cannot be on the midline, and the *input* muscle force is lower. Thus, the *sum* of the bite force and the two joint forces must also be lower, because they are reactions to the lower muscle force.

At the same time, **B4** is closer to the muscle resultant at **R′** and so the leverage of the jaw mechanism is better than it is at more anterior bite points. The interaction of both of these effects, less input muscle force and better leverage, produces a bite force at **B4** that is still substantial but not as large as might be expected from a study of the two-dimensional case.

Another region along the jaw extends from **M** to **J** (Fig. 2.2) – that is, from the location of the lateral resultant muscle force back to the jaw joint. If the bite point were at **M**, the resultant must also be at **M** if the jaw is to be stable. In that case the resultant lies at one angle, or vertex, of the triangle-of-support and there are zero forces at the other two vertices. For all bite points behind **M**, the resultant will always be outside the triangle-of-support. Thus, the jaw cannot be stable if the bite point is any farther back than **M**. The hypothetical bite point **B5** illustrates this condition; the triangle-of-support (**JB5J**) lies behind the point **M** that is essentially the most posterior location of the muscle resultant force. Joint dislocation and ligament endangerment occur because the muscle force lies outside the triangle-of-support and the joint ligaments are stressed.

[Recall that the location of **M** can move forward or backward by a small amount. This amount depends on the activity of the various parts of the jaw muscles. However, relative to the amount of mediolateral movement, the anteroposterior movement is fairly small and is not considered.]

Therefore, the entire length of the jaw, from the joint to the anterior end, can be divided into three regions. The two anterior regions have different bite force regimes and biting cannot take place at all in the posterior region. The most anterior region extends from the anterior end of the jaw back to point **B3**. Point **B3** is defined by a line beginning at the joint on the opposite side of the head. This line passes through point **R** on the midline that is located at the front of the posterior third of the jaw. (Point **R** is 30% of the way along the jaw, rather than about a third, as will be demonstrated below.) **B3** is the intersection of this line and a line extending along the *biting* tooth row from the joint to the incisor (**JB1**). Typically the premolars lie at the rear of this region. From **B3** to point **M** is the second region, and typically, but not always, only the molars are found here. From **M** back to the jaw joint is the third region, and it contains no teeth. In 1993 Spencer and Demes called these three regions, Regions I, II, and III where Region I is the most anterior (Fig. 2.2).

In Region I, (**B1** to **B3**) the resultant muscle force can be located on the midline, where it will have the largest possible magnitude (assuming maximum muscle force). (The resultant could also have a lower value if it is assumed that all or some of the muscles could be less active, although in these models maximum bite force, on at least one side, is a given.) If the muscle forces at **M** and **M** are not equal, the resultant will generally lie some distance toward the biting side, where the muscle force on that side is at or closer to the maximum.

Regardless of the location of the resultant (from **R** to **M** on the biting side) it will always be inside the triangle-of-support, and so the jaw will virtually always be stable. Anterior bite points will have lower bite forces than more posterior bite points because those farther forward are farther from the joints and the muscle force, and so the leverage is not as good. Locating the muscle force on the midline is almost always possible, because this force is almost always inside the triangle-of-support. Thus the largest possible muscle forces are possible on that part of the tooth row that extends from the incisors and usually includes the premolars (**B1** to **B3**).

For any single bite point in Region II (**B3** to **M**), where the molars are often located, not all muscle resultant locations are possible (Fig. 2.2). More posterior bite points *require* more lateral positions of the muscle resultant in the direction of the biting side. Lateral positions of the resultant are due to lower muscle forces on the non-biting side; maximum total muscle force is not available. Bite forces in this region must therefore be lower than an extrapolation from the bite forces in Region I would suggest. Nevertheless, they are not overly low because the leverage of the lever system is better at the rear of the jaw.

Finally, in Region III (**M** to **J**) the resultant is always outside any of the possible triangles-of-support. The jaw cannot be stabilized; the biting side jaw joint will always tend to be disarticulated, thus stressing the joint ligaments. Teeth are not found in this region. Recall that disarticulation of a jaw joint is a concern because of large tensile forces that will endanger the joint capsule and its ligaments and there are no anatomical structures present to prevent strong disarticulation.

At this point, the analysis dealing with the ideal combination of bite force and tooth row length in the two-dimensional case can be compared with the three-dimensional situation. In the two-dimensional case, all of the available muscle force was used for all the bite points. The tooth row, including the molar teeth, filled the entire region in front of the muscle force. In the three-dimensional case, all of the available muscle force can be used only in Region I. This region typically does not include the molars. Region I is thus shorter than the tooth row in the two-dimensional case. Different anteroposterior locations of the muscle force (different *hypothetical* anteroposterior locations of **R** in Figure 2.2) give different curves when bite force is plotted against location along the tooth row. The same was the case in the two-dimensional model (cf. Fig. 1.4). The largest area under the curve (the largest sum of bite forces) is found to be evident when the muscle force is positioned at 30% of jaw length from the jaw joints, measured along the midline in the three-dimensional case (see Box 2.1 below). When the area under the curve in Region II is included in the analysis, the total area remains largest when **R** is located at the 30% position.

In Figure 2.2, the distance along the midline from the jaw joints to **R** is thus three units, whereas the distance from **R** to **B1** is seven units. Therefore, **R** is expected to be located at location 3 along a ten-unit-long jaw where measurements are taken along the midline and perpendicular to the vector. This will insure that the largest amount of potential bite force will result regardless of the magnitude of the muscle force that is put into the jaw mechanism. See Box 2.1 for details of the calculations relating to the above.

2.6 A typical jaw

Figures 1.1 and 1.2 in Chapter 1 can now be modified in a few important ways. First, when measured perpendicular to the muscle vector the length of the jaw is ten units. The distance ***d***, from the jaw joint to the muscle force, can be adjusted from one unit to three units. These changes position the muscle resultant force at 30% of the way along the jaw, as suggested by the results of the analyses above.

In Figure 2.2 teeth are located from the anterior end at **B1** to the location equivalent to **M**, so that the entire tooth row extends from **B1** to **M**. As in the previous diagrams (Figs 1.1 and 1.2), measurements are taken *perpendicular* to the arrow representing the muscle force. These modifications produce a jaw with a tooth row that is approximately 70% of total *jaw* length. The length of the edentulous region behind the muscle force is 30% of jaw length. This is true because that is the location of the muscle force vector, and the teeth are anterior to this force. These relationships are seen in most real mammals, even though the location of the muscle vector must be estimated (see Chapter 3 for a more detailed discussion).

Plotting bite force against location along the *two*-dimensional jaw when the muscle force has a magnitude of one unit and is positioned at location 3 produces the curve in Figure 1.9c. Note that in the *three*-dimensional case, if the grand resultant muscle force is equal to one unit of force and is located on the midline, the lateral resultants are each 0.5 units. In Figure 2.2 each of the many bite forces in the region

Box 2.1

In the two-dimensional case the jaw is modeled as a single straight line. The muscle resultant force of one unit is located r units from the jaw joint and the distance to any bite point is b. Therefore, $B = r/b$ is the equation of the curve for bite force against location along the jaw, where B is the magnitude of the bite force (cf. Fig. 1.9c). Integrating this equation gives $r \log_e b$. For each location of the muscle force (e.g., $r = 2, 3, 4$, etc.) the expression $r \log_e 10/b$ evaluates the integral from the posterior end of the tooth row, where $b = r$, to the anterior end of the jaw, where $b = 10$. When the muscle vector is located at approximately 3.7 units from the jaw joint in a ten-unit-long jaw ($r = 3.7$), the integral has the highest value.

In the three-dimensional case the resultant force is located on the midline of a triangular jaw, r units from the level of the jaw joints. The line from the joint on the non-chewing side, through the location of the muscle resultant on the midline (r units from the joints) intersects the posterior end of the *laterally* positioned tooth row at b (cf. line **JRB3** in Fig. 2.2). This intersection is the posterior end of Region I. In terms of r and the anterior end of the jaw (10), analytic (coordinate) geometry indicates that this posterior end of the tooth row is $20r/(r + 10)$. The expression $r \log_e (r + 10)/2r$ evaluates the integral $r \log_e b$ from the posterior end of the tooth row ($20r/(r + 10)$) to the anterior end at 10. The highest value of the integral is evident when the muscle vector is located at 3.0 units from the jaw joints ($r = 3.0$) or 30% of the way along the jaw. The area under the curve for the region from **B3** to **M** (Region II) can also be measured (cf. Fig. 2.2). The best location for r remains at 3.0 when this area is included. For a fuller discussion, refer to Greaves (1988a) in the References section.

from **B1** to **B3** can be calculated by drawing a line from the bite point through the muscle resultant location on the midline at **R** and ending at its intersection with the line joining the jaw joints (cf. the lines from **B1**, through **R**, to line **JJ** or from **B3**, through **R**, to **J**). These lines represent the jaw levers, and distances between these three points (e.g. **JR** and **RB3**) are used to calculate the bite force, as was done in the two-dimensional case.

[In another approach, which amounts to the same thing, the jaw joints, the muscle resultant force, and the bite points of interest are projected onto the midline of the jaw. The resulting distances, along the midline, can then be used to calculate the bite forces (because all lines intersecting three parallel lines are divided in the same ratio).]

A bite force of approximately 0.65 units is found at bite point **B3**. The lowest bite force is found at **B1** and is 0.3 units. All the bite-force values from **B1** to **B3** have the same magnitudes as in the two-dimensional case. However, note that the distance from **B1** to **B3** is only part of the complete tooth row.

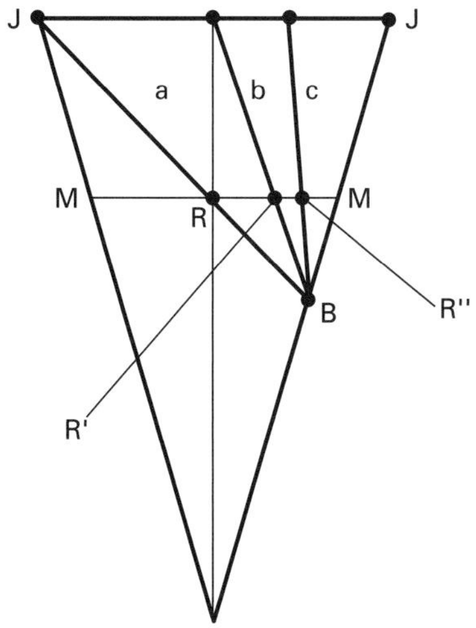

Figure 2.3. Bite force at B is highest when the jaw lever is line a and the resultant muscle force (R) is on the midline with a magnitude of one unit. The bite force at B is lower if the jaw lever is line b and the resultant muscle force (R′) is lateral to the midline, where its magnitude is less than one unit. The bite force is lower still if the jaw lever is line c because the resultant muscle force (R″) is even lower.

When calculating the bite forces in the molar region, from **B3** to **M**, the decreasing magnitude of the muscle force must be taken into account (cf. Fig. 2.2). The muscle force at the midline at **R** is one unit, but it is less than one unit at points lateral to **R** at, say, **R′**. The resultant muscle force must be more lateral and lower in magnitude for each more posterior bite point if it is to remain within the triangle-of-support as described above. In these cases only some of the muscle force from the non-biting side can be used.

The calculated bite forces from **B1** to **B3** are *maximum* forces because all of the available muscle force is used (0.5 on each side). To calculate the *maximum* bite forces from **B3** to **M**, a line from the bite point in question to the jaw joint on the opposite or non-chewing side is constructed (Fig. 2.2). For example, if **B4** is the bite point, a line from this point to the joint on the opposite side intersects line **MM** at **R′** (Fig. 2.2).

[A number of lines that begin at any bite point and intersect line **MM** could be constructed, as in Figure 2.3. Drawing the jaw lever to the joint on the non-biting side insures that the resultant force is as high as possible, because this lever intersects line **MM** as close as possible to the midline, where the muscle force is one unit. For example, consider the three hypothetical levers (**a**, **b**, and **c**) extending from the bite point **B** in Figure 2.3. If the lever is line **a**, it passes through point **R** on the midline and therefore the muscle force is as large as possible (one unit). If the lever is line **b** or **c**, the resultant force is shifted to the side (to **R′** and **R″**, respectively) and is therefore lower than the muscle force at **R** on the midline. In these latter cases the

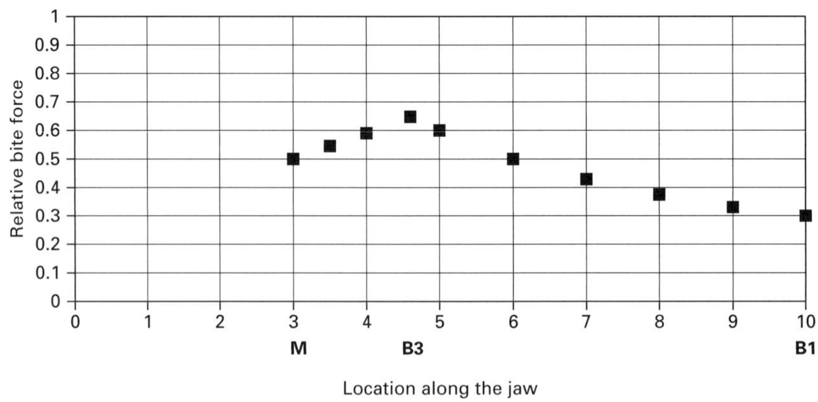

Figure 2.4. A plot of maximum relative bite forces against location along the jaw for the triangular jaw diagrammed in Figure 2.2. The magnitude of the muscle force on each side is 0.5 units.

bite force at **B** is lower because the muscle input force is lower. Note that lines **a**, **b**, and **c** are divided in same ratio by line **MM**.]

When biting at **B4** (Figure 2.2) and given that the muscle force on the biting side is 0.5, the problem is to determine the magnitude of the muscle force on the non-biting side such that the grand resultant is positioned at **R′**. The simplest procedure is to determine the *ratio* of the distances from **R′** to the **M** on each side. The muscle forces at each **M** are in the same ratio as these distances, with the larger force paired with the shorter distance. As the muscle force at **M** on the biting side is 0.5, the force at **M** on the non-biting side can be calculated using simple lever mechanics as was done above. The grand resultant in this case is simply the sum of these two forces at **M** and **M** on each side.

Once the magnitude of the grand resultant is known, the distances from **B4** to **J** and from **R′** to **J** can be measured (Fig. 2.2). The bite force at **B4** can then be calculated using elementary lever mechanics. The distance from **J** to **R′** is the input lever and the distance from **J** to **B4** is the output lever. The bite force at **B4** is found by using these distances together with the calculated grand resultant (at **R′**). Any bite force at any point between **B3** and **M** can be calculated in this way.

A plot of bite force along the tooth row from **B1** to **M** is shown in Figure 2.4. The curve from **B1** to **B3** (Region I) is the same as the curve in the two-dimensional case (cf. Fig. 1.9c). The curve from **B3** to **M** (Region II) in this posterior region of the tooth row is not the same as in the two-dimensional case. Here it is a line that slopes down toward the rear of the jaw instead of rising to very high values as it does in the two-dimensional case (cf. Figs. 1.9c and 2.4).

The curve from **B3** to **M** in Figure 2.4 slopes down *in this particular case* because the jaw was modeled as a triangle, in which the tooth rows form a V and therefore the rows strongly converge toward the front of the jaw. However, in other mammals, with differently shaped dental arches, the graph of bite force in Region II often falls

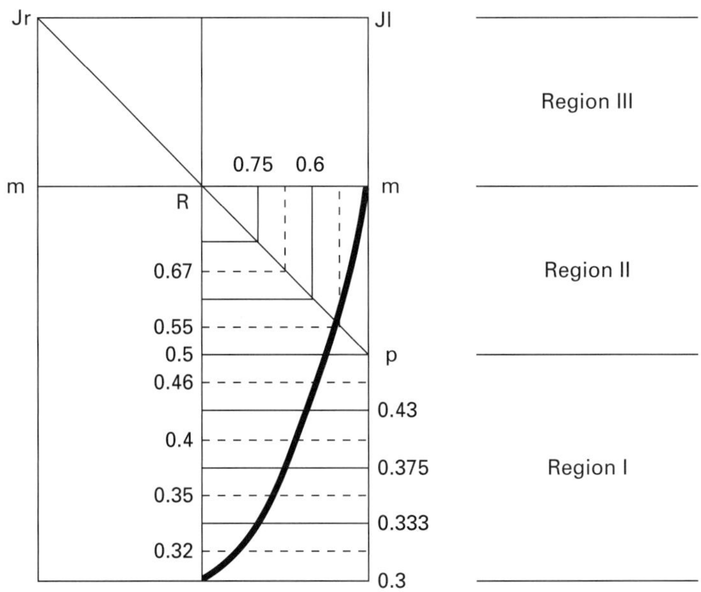

Figure 2.5. A "contour" map of bite forces. After Greaves, W. S. (2002). Modeling the distance between the molar tooth rows in mammals. *Canadian Journal of Zoology*, **80**, 388–393.

more gently, rises slightly, or is essentially horizontal. If the teeth in this posterior region are farther to the side (relative to the muscle force) and converge less strongly (as in humans), the bite force at the most posterior tooth is only slightly lower than the more anterior teeth in Region II. If the posterior teeth diverge toward the front of the jaw (as in some rodents) the bite force at the posterior tooth is slightly higher than that at the more anterior molar teeth. If the teeth on each side of the head in Region II are essentially parallel (as in many antelopes), the bite force at all of the teeth in this region is approximately the same.

Thus, the shape of the dental arch and the location of the muscle forces at **M** and **M** have an effect on the magnitude of the bite forces in the posterior region of the tooth row. Some researchers have suggested that there also appears to be a relationship between tooth size and bite force; teeth with higher bite forces are typically larger in real animals.

2.7 A contour map of the bite forces

As indicated, the tooth rows do not always form a triangle, but could form, say, a U-shaped dental arcade. Therefore, a "contour map" indicating relative bite force for any jaw shape would be of some interest. Figure 2.5 is such a map. It is a plan view of all locations where teeth would reasonably be found in different mammals (cf. Fig. 2.1). The "contour lines", or isopleths, in this case are isodynes that indicate the loci of points with equal bite force. Isodynes representing bite forces from

0.30 to 0.75 are included. In this diagram isodynes with low magnitudes are straight, horizontal lines, whereas those with higher magnitudes are reversed Ls that have straight, horizontal, and vertical limbs. A line representing almost any tooth row (for example, the heavy curved line in Figure 2.5) can be superimposed on this map. Like a line drawn on a topographic map, where the altitude at each intersection of a contour can be read off, bite forces can be read at each intersection of an isodyne with a line representing a tooth row.

The oblique line from **Jr** through **R** to **p** separates Region I (on the left) from Region II (on the right). Note that the *muscle* forces on each side (at **m**) are directly in line with the jaw joints (from front to back) and therefore are more laterally positioned than they were in Figures 2.1 or 2.2, where these locations were positioned along a triangular jaw. To simplify the diagram, only the isodynes on the left side of the jaw are shown.

The isodynes in Region I indicate that forces at lateral teeth are equal to those at teeth closer to the midline. As seen previously, bite forces are low at the front of a jaw and higher at the rear. The molars are typically located to the right of the oblique line in Region II. The isodynes in this region are vertical, indicating that bite forces are lower at the side of the jaw and increase toward the midline. A molar row in Region II that is parallel to the midline has bite forces that are approximately equal at all these teeth.

In Figure 2.5 the highest input muscle forces are assumed (0.5 units of force on each side at **m** and **m**, resulting in one unit of force at the midline of the jaw (at **R**)). When biting at the central incisor the jaw lever is the midline of the jaw. The midline extends from the level of the joints to the anterior end of the jaw and is divided in the ratio of 3:7 by **R**. The force at the incisor is 0.3 units of force and the forces at the joints must sum to 0.7 units of force. The midline divides the distance between the joints in half. The ratio of the distances to each joint is equal to the ratio of the forces at these joints. Thus, each joint resists 0.35 units of force. The input muscle force of one unit is balanced by the sum of the reaction forces at the joints and the bite point (0.35 + 0.35 + 0.30 = 1).

If the bite point is moved from the central incisor and is now located at the lower right corner of the diagram, where the bite force is still 0.3, the jaw lever extends from that point through the muscle resultant at **R** to a point between the two joints. This point of intersection is closer to **Jr** and farther from **Jl**. The line from the bite point, through **R**, to the line connecting the joints is longer than the midline, but is still divided in a 3:7 ratio by **R**, so that the bite force remains 0.30.

The ratio of the distances from the intersection of the jaw lever to each of the jaw joints in this second example is no longer one to one; the intersection of the jaw lever with the line connecting the joints is now closer to **Jr**. Therefore, the force at **Jr** is higher than the force at **Jl**. The joint forces still sum to 0.7. The force at each joint can be calculated by noting the intersection of the jaw lever with the line that extends from one joint to the other. The joint forces are in the same *ratio* as the distances from the intersection of the jaw lever to each joint. The higher force is paired with the shorter distance and the lower force with the longer distance.

The joint forces for any bite point can be calculated in a similar fashion. The line from the bite point through point **R** will intersect the line between the joints. Subtracting the bite force from one unit of muscle force gives the *total* joint force. Determining the ratio of the distances from the intersection of the jaw lever to each joint allows the calculation of the individual joint forces. This will be true for all bite points in Region I that are to the left of the diagonal (**JrRp**).

For bite points on this diagonal, the jaw lever intersects the right joint **Jr**. In these cases the joint force at the left joint **Jl** is technically zero. The jaw lever for bite points to the *right* of the diagonal does not pass through **R**, but for the largest possible muscle force the jaw lever must be as close to **R** as possible (cf. Fig. 2.3). This simply means that the jaw lever in each of these cases will intersect joint **Jr**, and therefore technically there will be no reaction force at joint **Jl**. Recall the previous mention of a safety factor so that **R** will likely be slightly farther toward the biting side and there will be some small force at joint **Jl** in real animals.

The heavy curved line in Figure 2.5 signifies a representative tooth row. At the approximate location of the third molar at the rear of Region II, this line intersects the isodyne with a value of 0.5 units. At the approximate level of the first molar at the anterior end of Region II, the 0.55 isodyne is intersected. The intersection of the curved line with the isodynes in Region I also indicates the magnitudes of the bite forces in this region.

To summarize the above, Figure 2.6 includes the plots for a unit muscle force, bite force, total joint force, and both joint forces, against location along the jaw for a triangular jaw such as in Figure 2.2. The total muscle force of one unit (open diamonds) begins to decrease behind location 4.6 because the muscle force on the non-biting side decreases in order to move the resultant muscle force inside the triangle-of-support. The relative bite force (filled squares) therefore also decreases just beyond this location (cf. Fig. 2.4). The working side joint force (small open circles) decreases steadily until it reaches zero just before location 4. The non-working, or balancing, side joint force (large open circles) remains relatively constant before decreasing more rapidly to zero to the rear of location 5. The plot for total joint force (small filled circles) indicates the sum of the working and balancing side joints (cf. Fig. 1.5).

A differently shaped tooth row naturally intersects the isodynes at different places (imagine a different heavy curved line in Figure 2.5). The bite forces in most cases will be the same in Region I, but will be slightly different in Region II. That is why the bite forces at the molars can be approximately equal, or increase or decrease somewhat toward the rear.

If the joints in Figure 2.5 were positioned more laterally (shifting line **Jrp**), Region I would be longer and Region II would be shorter, measuring anteroposteriorly. Nevertheless, because the lever system is fixed, the magnitude of the bite forces depends only upon the masses of the jaw muscles.

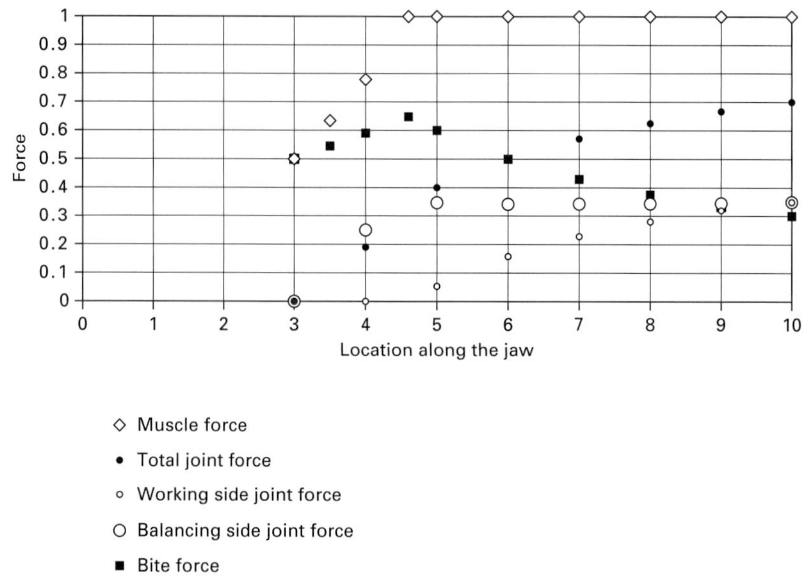

Figure 2.6. Plots that summarize the various relative forces when biting takes place at different points along the jaw when the bite force is located at location 3. Recall that there are no teeth behind location 3. The relative forces were calculated for a triangular jaw as in Figure 2.2. Open diamonds = muscle force; small filled circles = total joint forces; open small circles = working side joint force; large open circles = balancing side joint force; filled squares = bite force. Compare this figure with Figure 2.4.

2.8 Location of the carnassial tooth

The carnassial tooth, when it is present in carnivores, is located such that it usually overlaps the halfway point along the entire jaw. Therefore, it is located near the anterior end of Region II (cf. Fig. 2.5). Here, fairly large bite forces are available in a jaw where the right and left tooth rows form a V. The carnassials are therefore reasonably close to the back of the jaw so that the distance between occluding carnassials (gape) is still great enough to accept fairly large food items. Thus the location of the carnassials, at the anterior end of Region II, allows for quite large bite forces at a point where the gape is still reasonably wide.

Usually only a single pair, or perhaps two pairs, of teeth will be working against a food item at any instant in carnivores. When the upper and lower carnassial teeth meet each other, like most upper and lower tooth pairs, they are typically the only teeth that are working against each other. Thus, bite force is concentrated in a small region.

Carnassial teeth are long anteroposteriorly, but narrow and blade-like mediolaterally. In lateral view the blades of carnassial teeth have a large V-shaped notch near the middle of the cutting edge. The two edges within each notch are worn to sharpness and wear against a pair of similarly sharpened edges on the carnassial tooth in the opposing jaw. The anterior and posterior pairs of edges move against each other somewhat like the blades of a pair of scissors, although the blades of scissors rotate around a pivot point, whereas the blades of carnassial teeth translate past each other. Nevertheless, the effect is similar. Food caught between the blades is cut at the point where the blades meet at an angle and begin to shear past each other, in much the same way as material is cut by scissors.

As is well known, the importance of occluding blades with V-shaped notches that approach each other can be appreciated more clearly by imagining the pair of scissors. Consider the blades of scissors when working against resistant items. In this case the item to be cut is often *pushed* by the blades instead of being cut. This effect can be demonstrated by trying to cut a wooden pencil with a pair of ordinary scissors; the pencil simply slides forward away from the point where the edges of the blades meet and the pencil is not cut. If two V-shaped notches approach each other they form an opening that is shaped like a small parallelogram. A food item, resistant or not, that is trapped within this opening may still slide away from either the anterior or posterior pair of edges as carnassials move toward each other. However, each pair of these edges or blades pushes the food toward the other pair as the parallelogram-shaped opening becomes smaller and smaller. The food item is trapped and eventually cut because it can no longer slide away from either cutting edge.

A good analogy is a pair of ring shears. Ring shears have curved blades instead of the straight blades of a pair of normal scissors. Therefore, the blades shear past each other at two points instead of one. These blades form an "eye"-shaped space (instead of a parallelogram) and they trap, and eventually cut, items inside this space. The V-shaped edges on the blades of a laboratory guillotine function more like carnassial teeth.

Given the importance of cutting food items, it is not entirely unexpected that teeth behind the carnassial tooth are sometimes lost over evolutionary time in some of the Carnivora. Nevertheless, the muscle force is still expected to be located at about 30% of the way along the jaw so as not to disturb the basic 3:7 ratio that is so important for the jaw mechanism. In the majority of cases, the resultant muscle force intersects the tooth row just behind the third molar. However, if some teeth behind the carnassial are missing, naturally there will be some distance between the intersection of the muscle force and the carnassial tooth. Thus, in some carnivores, the resultant vector of muscle force does not intersect the tooth row just behind the most posterior tooth (because this tooth is not the third molar), but at some short distance to the rear of that point.

2.9 The mediolateral location of the tooth row

2.9.1 The sum of the bite forces

The approach taken to the analysis of the masticatory apparatus up to this point has been to determine the location of the muscle force such that the sum of the bite forces is maximized. A similar approach is employed when examining the location of the tooth row relative to the midline of the jaw when the anteroposterior location of the muscle force is fixed in position, where the muscle force divides the jaw in a 3:7 ratio.

The cheek tooth row is sometimes located far from the midline, more or less in line with the lateral edge of the jaw joints as in antelopes and related forms, and in animals like the somewhat similar horse. However, typically in mammals the tooth row is closer to the midline of the jaw. The molars at the rear of the tooth row are often about halfway from the midline to the lateral edge of the jaw joint. This is true in those cases in which the tooth rows are almost parallel to each other (as in many primates). It is approximately true even in those more common cases where the teeth converge toward the front, producing triangular jaws (as in carnivores like cats and dogs). The following simple analysis, like those used in previous sections, suggests that maximizing the sum of the bite forces along a tooth row requires that rows be located at, or near, this halfway position.

Figure 2.5 indicates that, at any given distance forward from the joints in Region I, the bite force at lateral and medial teeth is the same. Thus, as far as bite force is concerned it does not matter whether the teeth are more laterally or more medially located. Incisors can be positioned at the midline, but more posterior teeth that are too close to the midline are precluded because of the presence of the tongue. In Region II, more medial teeth exert higher bite forces so, as a first approximation, teeth are expected to be as close to the midline as possible. Accepting that tooth rows often form continuous lines from front to back, molars as close to the midline as possible (for the highest bite force) are expected to determine the location of the entire tooth row, because the mediolateral location of teeth in Region I has no effect on bite force.

Bite forces at teeth all along a tooth row can be determined, and these forces can be plotted on a graph. The area under the curve, which also represents the sum of the bite forces, can then be determined. The muscle force divides the jaw in a 3:7 ratio (anteroposteriorly) when the sum is largest. The same approach can be applied to a number of different *hypothetical* tooth rows, in the same animal, that are more laterally or more medially situated with the muscle force always at 30% of jaw length. Tooth rows might be V- or U-shaped, but their mediolateral location is also important. The row with the highest sum of the bite forces, for a number of hypothetical rows with different mediolateral locations in the same animal, is presumed to be in the "best" location. This best location turns out to be where tooth rows in many mammals are located.

2.9.2 The length-to-width ratio of the jaw

Using the above approach, a best location for a tooth row in a given animal can be chosen from a series of hypothetical rows that are located more medially or more laterally. However, different real jaws have different length to width ratios and the best location is different for relatively wide and relatively narrow jaws. The best location for the tooth row in a wide jaw is farther from the midline than in a narrow jaw. If the distance between the jaw joints is 60% of jaw length, the jaw is relatively long and narrow. The distance between the molars in this case will be about half the distance between the joints. If the distance between the jaw joints is 80% of jaw length, the jaw is relatively short and wide. Here the distance between the molars is closer to about 62% of the distance between the joints.

When a large number of different mammals are examined, the distance between the two jaw *joints* is between 60 and 80% of jaw length when length is measured along the midline. Thus, recalling the previous paragraph, the distance from molar to molar is expected to be between 50 and 62% of the distance from one joint to the other. This is exactly what is found in many cases. Mark Spencer measured more than 700 anthropoid primates in his 1995 dissertation. In almost all cases the distance from one molar row to the other was between 50 and 60% of the distance between the jaw joints. Unfortunately, Spencer's dataset did not allow the length-to-width ratio of the jaws to be determined. Thus, a definitive statement cannot be made using these particular data.

Mammalian jaw lever systems have the same 3:7 ratio in an anteroposterior direction, because this relationship gives the largest total bite force. As the lever system is fixed at the most efficient configuration, larger muscle forces are required if larger bite forces are needed, because the lever system is not expected to change. Furthermore, more musculature can be packed into jaws that are wider. Thus, the width-to-length ratio of the jaw is a crucial measure. As a general rule, in animals with similar jaw lengths, those with wider jaws will have higher bite forces.

For example, a bobcat is a larger animal than a gray fox, but the two animals have jaws of essentially the same length. However, the jaw of the bobcat is much wider, and thus there is more room for the obviously larger jaw muscles. Larger muscles mean higher bite forces. The same is true of a mountain lion, with a wide jaw, when compared with a medium-sized domestic dog, with a narrow jaw, when they both have the same jaw length.

Finally, the jaws of most antelope and their kin (more modern selenodont artiodactyls) have become longer compared with more primitive members of this group and have lost the upper incisors, usually the canine, and some premolars. Tooth size has not changed appreciably, so a very long diastema, or region without teeth, is present in the lower jaw between the canine and premolars. This jaw lengthening might be related to producing a longer and narrower snout that improves the ability to choose particular parts of plants. However, this lengthening of the jaws also lengthens Region II. Some premolars are now located in Region II and exert higher bite forces than they did when they were located in Region I (cf. Fig. 2.5). They now have the higher bite forces typical of the molar teeth. Moreover, a small premolar with higher bite force, because it is now located inside Region II, is expected to enlarge, approaching the size of the molars.

A second change in some artiodactyls also increased the length of Region II. The cheek teeth are essentially in line with the jaw joints and thus are located in a relatively lateral position. Region II is longer laterally and shorter medially (Fig. 2.5). Two additional premolars are located in a longer Region II. Thus, jaw lengthening and more laterally positioned cheek teeth have effectively increased the length of Region II, resulting in the modern selenodont artiodactyl condition, in which in addition to the molars three fairly large premolars have higher bite forces because they are found in Region II.

More lateral molar teeth have lower bite forces (Fig. 2.5). However, six large molar-like teeth (three molars and three premolars) with slightly lower bite forces are better than only three molars with slightly larger bite forces. This is true especially since premolars, originally located outside of Region II, probably did not function exactly like molars.

There are no teeth in the very long diastema in selenodont artiodactyls and similar animals such as horses. Therefore, there are no bite forces in this region and the sum of the bite forces along the entire tooth row is smaller than expected. Except for the cropping teeth at the extreme anterior end of the lower jaw, all the other teeth, with few real exceptions, are located in Region II. Thus, concentrating our analysis on the cheek teeth seems reasonable.

The increased bite force at premolars, together with the longer and narrower snout, that may have improved cropping ability, are probably important parts of the reason for the selection of lengthened jaws in these animals. See further discussion in section 3.5.1 in Chapter 3.

2.10 Occlusion of the tooth rows

The location of the glenoid, or upper part of the jaw joint, on the skull has been mentioned above, but the location of the condyle, the lower part of the joint that is located on the lower jaw, has not been considered in any detail. Generally, the glenoid is some distance above the level of the upper tooth row as described above. However in the lower jaw, the distance from the level of the lower tooth row to the condyle is more variable. The condyle is often well above the tooth row, but is just as often close to the level of the row. Only rarely is it below this level.

The two distances, from the level of the upper tooth row to the glenoid and from the level of the lower tooth row to the condyle, determine how the upper and lower tooth rows meet at occlusion. In lateral view, if these distances from the joint to the respective tooth rows are equal and both the glenoid and condyle are above, below, or on the same level as the tooth row, all the teeth will occlude with their opposites at the same time (Fig. 2.7, upper panel). That is, the upper and lower tooth rows and a point representing the joint can be superimposed.

By contrast, the upper and lower tooth rows do not meet each other like this if the distances from the rows to the joint are not the same. In this case, the upper and lower tooth rows will meet at some angle to each other in *lateral* view. This will also be true if the glenoid is above and the condyle is below the tooth row, or *vice versa* regardless of whether the distances are equal or not (Fig. 2.7, lower panel). In this case, the two tooth rows and the joint cannot be superimposed.

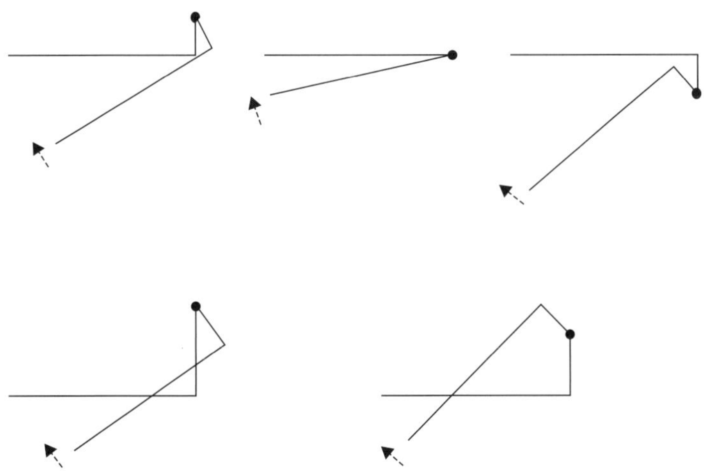

Figure 2.7. Stick diagrams of upper and lower jaws. The filled circles represent the jaw joints. Upper, simultaneous occlusion occurs when the distance from the joint to the upper tooth row is equal to the distance from the joint to the lower tooth row (the upper and lower tooth rows and a point representing the joint can be superimposed). Lower, tooth rows meet at an angle when: the distance from the joint to the upper tooth row is *not* equal to the distance from the joint to the lower tooth row (the tooth rows and the joint cannot be superimposed); one tooth row is above the joint and the other is below.

There have been previous discussions about how tooth rows meet each other at occlusion. Some researchers have suggested that the rows form an angle with each other, like the blades of scissors when jaw joints are on the same level as the tooth rows. Joints not on the same level were thought to allow all the teeth to occlude at the same time. However, a close examination of a pair of typical scissors clearly demonstrates that the rivet and the edges of the blades are on very different levels and so shows clearly that the actual case is just the opposite of the above views and somewhat more involved. This is an example of a problem mentioned in the preface. Unexamined acceptance of incorrect ideas can delay a more correct analysis for some time (cf. Fig. 2.7).

Consider the cases in the upper panel of Figure 2.7. Often in these situations, even though all the teeth meet simultaneously in *lateral* view, the tooth rows meet at some angle when viewed from *above*. Therefore, only one or a few teeth will meet their opposites at any one time in sequence from back to front. Selenodont artiodactyls and horses are good examples.

Typically in carnivores the glenoid on the skull is higher above the upper tooth row than the condyle is above the lower row. This means that, in lateral view, the upper and lower tooth rows will meet each other at an angle when the jaws close and a limited number of teeth will work against their opposites at any given time (Fig. 2.7, lower panel). The *entire* upper and lower jaws, with their tooth rows, superficially will look like the blades of normal scissors when they are closing.

2.11 Tooth-wear patterns

A tooth-wear pattern that is present on the teeth of many mammals is especially obvious on the teeth of selenodont artiodactyls. Food and the opposing teeth wear away the extremely hard and relatively thin layer of enamel on the surface of the crescent-shaped cusps of the teeth in these animals. This wear exposes the dentine that makes up most of the body of the tooth. Dentine is very hard but is softer than the even harder enamel. Therefore, the dentine wears away somewhat faster than the enamel, once the enamel is worn away enough to expose the dentine. The result is a wear facet, where the surface is made up of a large dentine region surrounded by a thin band of enamel. The dentine, being somewhat softer, wears to a greater extent and somewhat *lower* than the surrounding enamel; the result resembles an enamel "bowl" almost filled to the brim with dentine. Closer examination of these facets indicates that on one side of the "bowl" the dentine and enamel are flush, or on the same level. On the opposite side, the dentine and enamel are not flush but rather form a distinct step, with the enamel higher than the dentine. This morphology can be interpreted as follows.

The enamel bands or ridges contact each other first as the teeth occlude. Both enamel and dentine are worn away as the teeth, or the food located between them, move across each other. As the enamel on one side of the tooth is encountered first, the dentine that immediately follows it is protected to some extent, as the occluding tooth wears away both the enamel and dentine. Thus, the enamel and the dentine surfaces are worn to essentially the same level, even though the dentine is softer.

On the opposite side of the "bowl" there is no enamel ridge to protect the softer dentine because the enamel is now located behind the dentine. Therefore, the softer dentine is worn somewhat lower than the enamel and a distinct step is formed. Thus, by noting the location of the flush surfaces of the enamel and dentine (leading edge) and the stepped surfaces (trailing edge) the direction of travel of the occluding tooth can be inferred. The occluding tooth wears against the flush surfaces first, followed by the stepped surfaces.

2.12 Summary

The triangle-of-support is a major feature of the three-dimensional jaw. The three vertices of the triangle are located at the two jaw joints and the teeth at the bite point. The movable bite point largely defines the shape of the triangle because the joints are essentially fixed in position. The resultant force of the jaw muscles must be located within this triangle if the lower jaw is to be stable. If the muscle force lies outside the triangle, one of the jaw joints is pulled apart or is disarticulated.

Maximum muscle force, for a given location, can be exerted when biting takes place at the anterior portion of the tooth row. However, at the rear of the tooth row the action line of the muscle force must be shifted to one side if it is to remain inside the triangle-of-support and thus avoid dislocation of one of the jaw joints. Lower activity of the muscles on one side of the head is required to shift the muscle force to the opposite side. This lesser activity necessarily reduces the total muscle force. Therefore, at this posterior region, which is often the molar region, lower output bite forces are evident because less muscle force was put into the system, even though the leverage of the system is better.

In a three-dimensional jaw the largest sum of the bite forces, when summed along the entire jaw, is evident when the resultant force of the jaw muscles is located at 30% of jaw length from the jaw joint. That is, in a ten-unit-long jaw, the distance from the joint to the line-of-action of the jaw muscle resultant is three units, whereas the distance from the resultant to the anterior end of the jaw is seven units. This 3:7 ratio seems to be common to most mammals. Differences in bite force then depend only on differences in muscle mass.

Simple calculations allow the construction of a map that indicates the relative bite forces at the expected locations of teeth in many mammals. Points with the same magnitude can be connected by "contour lines" called isodynes. A line representing the tooth row can then be superimposed on this map. Bite forces all along the lines representing the "tooth rows" then can be read off such a map.

The sum of the bite forces for a number of hypothetical tooth rows, some more medial and some more lateral, can be calculated. Imagine superimposing a series of lines representing more medial and more lateral tooth rows on the isodyne map. After controlling for the length to width ratio of the jaws, the tooth row with the largest bite force sum tends to be located in essentially the same place in many mammals.

In a lateral view, the upper and lower tooth rows can either meet simultaneously all along their length or meet at some angle to each other. In general, if the tooth rows occlude simultaneously, the joint and all the teeth in one jaw can be superimposed on the joint and all the teeth in the other jaw. In all other cases, where superimposition is impossible, the tooth rows meet at some angle to each other.

The surface topography of a wear facet, which exposes the dentine on a tooth, can be used to determine the relative motion between two occluding teeth. On the side of the tooth that meets the occluding tooth first, the enamel and dentine form a flush, or smooth, surface. On the opposite side of the tooth the enamel and dentine form a stepped surface, where the dentine is lower than the enamel. The opposing tooth meets the flush surface first before the stepped surface. Therefore, the direction in which the occluding tooth is moving can be determined.

3 Vector inclination and joint location

The line-of-action of the resultant vector of jaw muscle force is inclined posteriorly in most orders of mammals. This inclination varies directly with the height of the jaw joint above the level of the upper tooth row. A very high joint is paired with a near vertical line-of-action. A low joint is associated with a vector with a strong posterior inclination.

A few orders of mammals contain species in which the rostral region of the jaw has elongated over evolutionary time. This elongation did not change the 3:7 ratio of the jaw regions. Concomitant changes in the masses of the jaw muscles shifted the inclination of the resultant muscle force vector anteriorly. This changed the moment arms of the vector, and the teeth, thus maintaining the ratio. Other than jaw lengthening, major changes in the jaws did not take place.

A study of jaw muscle size and third molar location provides an estimate of the anteroposterior location of the resultant vector of jaw muscle force. This estimate agrees with the analyses described above as well as with estimates derived from dissections of the muscles.

3.1 The inclination of the resultant vector of jaw muscle force

Both the inclination of the resultant vector of jaw muscle force and the location of the jaw joint on the skull can be related to the upper tooth row. First, that part of the jaw joint that is located on the skull typically lies above the level of the tooth row in mammals. Second, the resultant vector of jaw muscle force is most often inclined posteriorly relative to the tooth row. Less often the anterior angle between the vector and a line estimating the tooth row is less than 90° and the vector is anteriorly inclined. The following analyses suggest that there are a small number of constraints on the jaw mechanism that necessitate a close relationship between the posterior inclination of the muscle force vector and the position of the cranial part of the jaw joint. When the vector is strongly inclined to the rear, the joint is low and close to the level of the tooth row. Higher joints are paired with more vertical, but still posteriorly inclined, vectors.

The jaw mechanism can be modeled using only four points. Three of these points are positioned at obvious landmarks on the skull. The fourth point is derived.

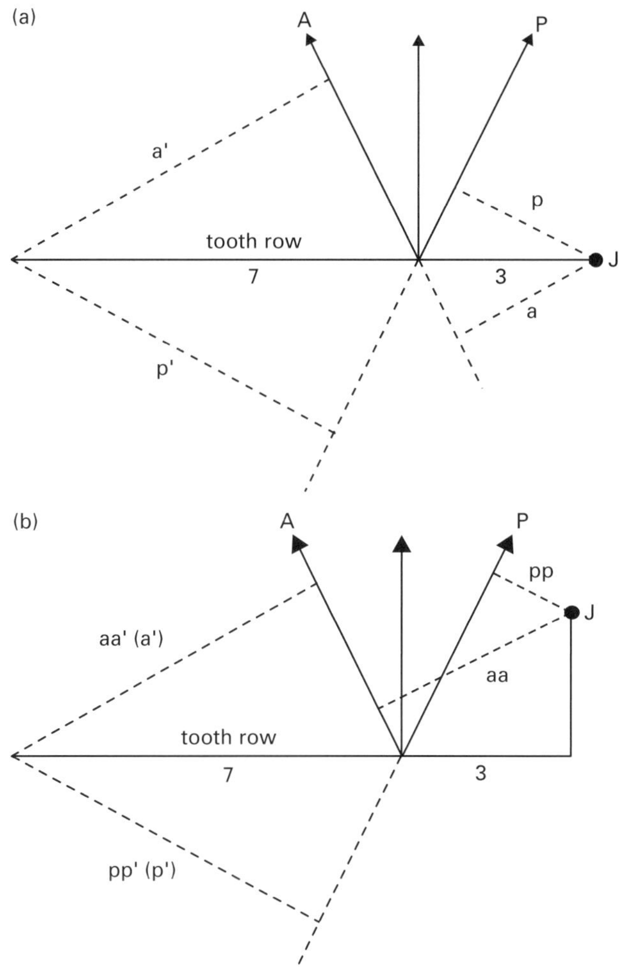

Figure 3.1. (a) The jaw joint (filled circle) is on the same level as the horizontal tooth row. (b) The jaw joint is located above the tooth row. Anterior (A), posterior (P), and vertical arrows represent resultant muscle forces. The lines labeled p and p′ in the upper panel are the *projections* of the posterior edentulous region of the jaw from the joint to the third molar and the anterior tooth-bearing region, onto lines that are perpendicular to the vector labeled P. The labels a and a′ indicate these same *projections* for the anterior vector labeled A. Corresponding projections in the lower panel are pp (for p), pp′ (for p′), aa (for a), and aa′ (for a′).

According to the model described below, the number of ways these four points can be related to one another is very limited. That is, only certain configurations of the four points satisfy a small number of constraints. A relatively small number of mammalian species have been examined closely so far, but they represent many of the major groups of mammals. In these cases the four points are configured as expected.

In spite of the close relationship between the *posterior* inclination of the muscle vector and the position of the joint relative to the level of the tooth row, some of the following analyses will consider either the vector inclination or the joint position in isolation. Later analyses will integrate these preliminary studies.

At the outset, recall that the *upper half* of the jaw joint, the glenoid cavity on the skull, is almost always positioned above the level of the *upper* tooth row in mammals. Rarely is it on the same level and even more rarely is it below the level of the tooth row, as may be the case in saber-toothed cats. The distance from the level of the *lower* tooth row to the *lower half* of the jaw joint, the condyle on the lower jaw, is more variable. The importance of these distances, in the skull and the lower jaw, is that they determine the way the upper and lower tooth rows meet each other in lateral view when the jaw closes. These relationships were examined previously in Chapter 2.

In earlier analyses (in Chapters 1 and 2), the resultant muscle vector and the bite and joint forces were represented as being vertical relative to a horizontal tooth row (cf. Fig. 3.1a, vertical arrow). All distances were measured perpendicular to the muscle vector and therefore *along* the jaw. This procedure simplified the arithmetic when using elementary lever mechanics. However, in most real mammals, the actual inclination of the resultant vector of muscle force, in lateral view, depends on the relative sizes of the three great jaw-closing muscles: temporalis, masseter, and medial pterygoid. The temporalis slopes up from the lower jaw to attach in the temporal fossa on the lateral side of the braincase. Thus it slopes toward the rear. The other two muscles, masseter and medial pterygoid, slope up and toward the front of the skull, because they attach at the rear of the jaw and more anteriorly on the skull.

When the temporalis is larger than the other two muscles taken together, the *resultant force* of the jaw muscles generally has a posterior inclination. It typically has an anterior inclination when the masseter and pterygoid form the dominant muscle group. Thus, the inclination of the muscle force varies because there is variation in the size of the individual muscles that close the jaw.

Most of the major mammalian groups have a resultant muscle force vector that is inclined toward the rear because they have a large temporalis muscle. Some have suggested that a large temporalis may resist anteriorly directed external forces on the lower jaw. For example, the large temporalis muscle in carnivores that struggle with their prey may resist large, anteriorly directed forces exerted on the lower canine teeth.

The temporalis tends to be relatively smaller in modern herbivores, in which the masseter and pterygoid are relatively larger. In these animals the resultant muscle force is usually inclined toward the front of the skull. The reason for an anterior inclination appears to be related to jaw lengthening, which in turn requires such an orientation. This relationship will be examined fully in section 3.5.1 below.

Thus, typically in mammals the resultant vector of jaw muscle force slants either anteriorly or, more often, posteriorly as it extends from the lower jaw up to the skull. In all cases the inclinations of the bite and joint forces are still parallel to the resultant muscle force because they are both forces that are present in reaction to the muscle force. Often, studies of the jaw mechanism consider *components* of the joint and bite forces that are perpendicular to joint surfaces or tooth facets. These components are often not parallel to the muscle force. Nevertheless, the bite and joint forces themselves remain parallel to the muscle force.

The analyses in Chapter 1 modeled the jaw as a single straight line, as in Figure 3.1a, with the jaw joint (filled circle) on one end and the teeth at the other. The largest total bite force is realized, in the *two-dimensional* case, when the vector is slightly greater than a third of the way along the jaw.

In the *three-dimensional* case, 30% of the way along the jaw is the best location. The vertical resultant vector of jaw muscle force forms a right angle with a ten-unit-long jaw and is located three units from the jaw joint (see the middle arrow in Figure 3.1a). Therefore the moment arm of the vector, as it pulls the lower jaw up, has a length of three units, and the moment arm of the most anterior tooth has a length of ten units. Thus, the ratio of the distances, from the vector to the joint and from the vector to the anterior end of the jaw, is 3:7. Again, considering the *two*-dimensional case, the vector is a little more than a third of the way along the jaw. Considering the jaw mechanism as a more appropriate *three*-dimensional structure, the largest total bite force is realized when the vector is *30%* of the way along the jaw.

In the simple case diagrammed in Figure 3.1a, as described in the following section, the inclination or orientation of the line-of-action of the resultant vector of muscle force has no effect on the mechanical advantage of the system; the leverage is exactly the same for *any* inclination of the muscle vector. This is *not* true if the jaw joint lies above (or below) the general level of the tooth row (Fig. 3.1b). In such cases, the mechanical advantage changes as the inclination of the vector changes.

3.2 Joint and tooth row on the same line

First consider the case in which the joint and the tooth row lie on the same line (Fig. 3.1a). In this diagram all the vectors intersect the tooth row of the lower jaw at the same point, just behind the third molar. The moment arm of the *vertical* vector has a length of three units (the perpendicular distance to the joint). A posterior vector (P) has a somewhat shorter moment arm, as indicated by a distance **p**, which is the *projection* of the distance of three units onto a line perpendicular to the posterior vector. Similarly, the perpendicular distance (**p**′) from the vector (P) to the anterior end of the jaw is shorter than seven units because it is also a projection. However, **p**:**p**′ = 3:7 because both of these distances are shorter by the same relative amounts. The absolute lengths of distances **p** and **p**′ are shorter than the three and seven units in the diagram, but their lengths are still in the ratio of 3:7. Thus, inclining the vector in any posterior direction does not change the mechanical advantage of the jaw lever system in the case where the jaw joint is on the *same* level as the tooth row.

The same is true for an anteriorly inclined vector (A) and for any other vector that is inclined away from the vertical in an anterior direction. The relevant projections in the case of vector A are indicated as **a** and **a**′ (Fig. 3.1a). For any anterior vector that intersects the jaw three units from the joint, regardless of its inclination, *projections* of the distances from the vector to the joint and from the vector to the anterior end of the jaw (e.g., **a** and **a**′), that are perpendicular to the vector, can be drawn. In all cases, distances equivalent to **a** and **a**′ will be in the ratio of 3:7 because these projections are reduced by the same relative amounts. The absolute distances from a vector to the joint (e.g., **a**), and to the end of the jaw (e.g., **a**′), change with the inclination, but the *ratio* of these distances remains 3:7, just as was the case with posterior vectors.

Thus the 3:7 ratio remains constant regardless of the inclination of the vector as long as the jaw joint is on the *same* level as the tooth row (Fig. 3.1a). The situation is very different if the jaw joint is located at some distance *above* (or below) the general level of the tooth row (Fig. 3.1b).

3.3 Joint above the tooth row

There is more than one reason for the evolution of the high jaw joint in mammals. One important reason derives from observations made by G. Becht in 1953. He noticed, as did others, that: (1) the jaw joints in a *frontal* view of a cow are positioned high above the tooth rows; and (2) the lines-of-action of the masseter and medial pterygoid muscles incline somewhat laterally and medially, respectively (cf. Fig. 3.2). He also noted that an occlusal plane can be extended from a low tooth row on one side of the head, up to the high jaw joint on the opposite side, with each plane forming one of the bars of an X in a frontal view (only the right occlusal plane (dashed line) is shown in Figure 3.2). However, Becht's key observation was that the action lines of the muscles *on both sides* intersect either occlusal plane at angles that produce components of force that move the jaw in only one direction. These components move the jaw from a lateral to a medial position and thus power the medially directed power stroke of the ungulate (in this case a cow) and by extension other mammals that chew in a similar way.

If the jaw joints are on the same level as the tooth rows, the occlusal planes are horizontal and coincident and do not form an X. In this case, the action lines of the muscles intersect the occlusal planes at angles that produce components of force that work against each other and therefore cancel each other to some extent. If the jaw joints are not situated *above* the tooth rows, two different occlusal planes are *not* defined and all the components of force do not pull the jaw in the same direction.

[The occlusal plane just described in the first paragraph includes a low bite point and two high jaw joints. The triangular plane defined by these points is steeply inclined. This actual occlusal plane is perpendicular to the anteriorly inclined muscle resultant in this animal. Note that the planes depicted in Figures 2.1 and 2.2 are also drawn perpendicular to the resultant.]

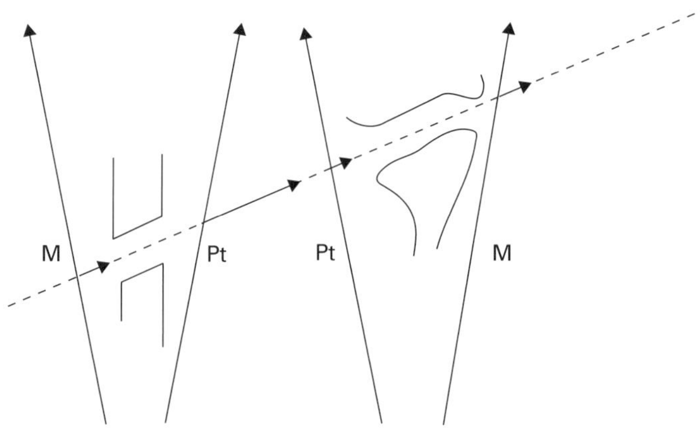

Figure 3.2. A diagrammatic frontal view of an antelope or horse showing the masseter (M) and medial pterygoid (Pt) muscle lines-of-action intersecting the right occlusal plane (dashed line). The muscles on both sides of the head produce components of force (short arrows) in the occlusal plane that move the teeth on the right side medially (toward the left side). The same four muscles intersect the left occlusal plane (not shown) to produce components of force that move the teeth on the left side medially (toward the right side).

Returning to our analysis, the posterior vector (P) in Figure 3.1b has the same inclination as the posterior vector (P) in Figure 3.1a. The projection of the rather long distance (not marked) from the joint to the third molar onto a line perpendicular to P (here labeled **pp**) is relatively short in this case. This is true even though the direct distance from the vector's intersection with the tooth row, just behind the third molar, up to the joint is fairly long. Moreover, the distance **pp** in Figure 3.1b is very obviously shorter than the distance **p** in Figure 3.1a. The distance **p** is the *projection* of the distance from the vector to the jaw joint when the joint is on the same level as the tooth row; distance **pp** is the equivalent projection when the jaw joint lies above the tooth row.

In Figure 3.1b the projection (**pp**′) from the vector to the anterior tooth, again measured perpendicular to the vector, is the same length as distance **p**′ that extends from the vector to the anterior end of the jaw in Figure 3.1a. The ratio between **pp** and **pp**′ (or **p**′) is therefore obviously not 3:7. In this case it is rather closer to about 2:8.5. When the joint is high, the *ratio* of the lengths of the projections of (1) the tooth row and (2) the distance from the joint to the molar, both onto a line perpendicular to a posterior vector, is different for every different posterior vector. For any *posterior* vector in this figure, none of these ratios is 3:7. In Figure 3.1b *only* the vertical vector divides the jaw in such a way that perpendicular projections are in the ratio of 3:7. Moreover, only in this particular case do the perpendiculars lie *along* the actual jaw.

In Figure 3.1b the projected distance (**aa**) from the jaw joint to the anterior vector (A) is longer than the equivalent distance in Figure 3.1a (distance **a**) because the high joint is in a different location relative to the vector's intersection with the tooth row. Nevertheless, the projection (**aa**′) in Figure 3.1b is equal to distance **a**′ in

Figure 3.1a. Therefore, if the ratio of **a** to **a′** in Figure 3.1a is 3:7, this cannot be true for the ratio of **aa** to **aa′**. The ratio of the two projections that are perpendicular to vector A (about 3.5:6.5), as well as the equivalent projections for any other *anterior* vector differs from 3:7.

Any vector, regardless of its inclination, in Figure 3.1b, can have projections that are in a 3:7 ratio, but in such cases the lengths of the jaw segments (and their projections), or the intersection of the vector with the diagrammed tooth row, have to be adjusted. For the cases diagrammed in Figure 3.1b and assuming the jaw joint is fixed in position, a 3:7 ratio requires that the anterior segment be shorter if the vector is posteriorly inclined and longer if it is anteriorly inclined. In Figure 3.1b **pp**:**pp′** = 2.3:7.7 (not 3:7) and **aa**:**aa′** = 3.5:6.5 (again not 3:7). Alternatively, if the *horizontal* length of the jaw in Figure 3.1b is not changed, vectors that are *not* vertical must intersect the jaw farther to the rear or farther forward in order to achieve a 3:7 ratio between the projections.

The jaw joint is typically located above the level of the *upper* tooth row in mammals. The resultant muscle vector is expected to be located such that the projections of the distances, to the joint and to the tooth row, are in the ratio of 3:7 to maximize the total bite force. The significance of the simple analyses just described is that, if the joint is high, the ratio of the projections to the vector can be determined if the locations of the jaw joint (glenoid cavity), the upper third molar, and the anterior upper incisor are known. If the joint is on the same level as the tooth row, all the projected ratios are the same regardless of the inclination of the vector.

In practice, the appropriate inclination of the vector is found by first determining the relative positions of the points representing the jaw joint (glenoid cavity), the rear of the upper third molar, and the anterior end of the upper jaw. After plotting the points on graph paper, a preliminary line is drawn through the point representing the rear of the third molar. The perpendicular distances from this preliminary line to the joint, and to the anterior of the jaw, are then measured. If the ratio of these two distances is 3:7, then the line as drawn is a good estimate of the best inclination of the vector according to the model. If the ratio of the distances is not 3:7, the inclination of the preliminary line can be changed by rotating it in the plane of the paper, around the point representing the molar, until a 3:7 ratio is achieved. This procedure determines the inclination of the vector as long as a third molar is present and the jaw joint is *not* on the line that extends along the tooth row. In cases in which the third molar has been lost over evolutionary time, the length of the missing tooth can be estimated. Naturally this latter procedure gives a less secure estimate of the inclination of the vector.

In any event, a reasonably good estimate of the inclination of the resultant vector of jaw muscle force can be determined in most mammals if the 3:7 ratio is accepted and the joint lies *above* (or in fewer cases, below) the level of the upper tooth row. When the joint is on the same level as the tooth row, *any* inclination of the vector gives exactly the *same* ratio, so a unique *inclination* cannot be determined.

Accepting the foregoing elementary analyses, the actual lengths of the distances from the vector's intersection to the joint and to the anterior end of the jaw, when measured along the jaw itself, are therefore *not* the critical measurements. The important distances are the lengths of the *projections* of each of these regions onto a line (or lines) *perpendicular* to the resultant vector of jaw muscle force. In most real mammalian jaws, where the muscle force is inclined either toward the back or front of the head, the length of the *tooth row* and the distance from the third molar to the jaw joint are not perpendicular to the muscle force. The actual lengths of these distances are typically longer than the projected lengths.

The relationship of 3:7 between the projections appears to be an almost universal condition in mammals and suggests that maximizing bite force, after insuring that the joint is not dislocated, is an important feature of the mammalian jaw mechanism. Recall that when a muscle vector divides a jaw in this way the total bite force along a tooth row is maximized. If the length of one of these projections were to change, the other would be expected to change in order to maintain the 3:7 ratio. Obviously, the inclination of the muscle force must either be known (as estimated by a careful dissection and weighing of the jaw muscles) or *estimated* in an animal (as described in the previous paragraphs) if reasonably correct *relative* bite and joint forces are to be calculated in any given case.

The ratio between these projections is not particularly obvious in mammals unless one actually looks for it. Most researchers examine more obvious features such as the length of the tooth row or the distance from the joint to the third molar. Few people seem to have concentrated on this line-of-action and how it is related to the jaw joint and various parts of the tooth row. Estimating the line-of-action of the *resultant* vector of jaw muscle force is somewhat tedious and is, after all, only an estimate. These may be the reasons that the 3:7 relationship has generally gone unnoticed. Yet once this ratio is observed it is seen to be the typical condition in mammals. Even rough estimates of the vector inclination demonstrate the relationship quite clearly.

3.4 Posteriorly inclined vectors

3.4.1 The direct distance from the joint to the third molar

If jaw mechanisms are examined with some care, an interesting difference is apparent between animals with a muscle vector whose line-of-action is anteriorly inclined and those with a vector with a posterior inclination. The major features can be seen in a lateral view (Fig. 3.3a). The jaw joint (glenoid cavity) is located at the end of the short limb of an L-shaped line that more accurately represents the upper jaw as well as many lower jaws. The jaw is L-shaped because the jaw joint on the skull (and often on the lower jaw) usually lies at some distance above the level of the horizontal limb of the L, which approximately passes along the tooth row.

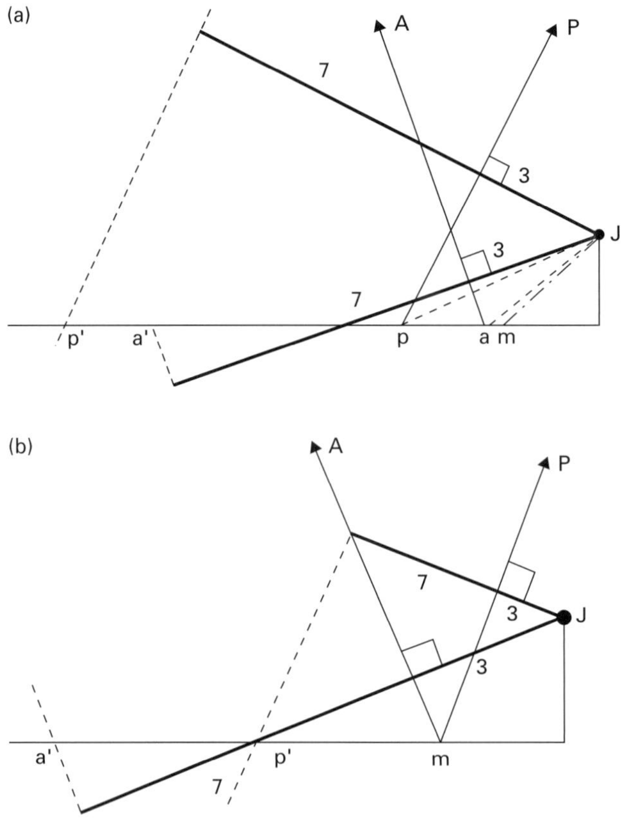

Figure 3.3. (a) A posterior (P) and anterior (A) vector included on the same jaw. The filled circle is the jaw joint. Note that the distance (Jp) from the jaw joint to the third molar at p is longer for a posterior vector than is the equivalent distance (Ja) for an anterior vector. The shortest possible distance from the joint to the third molar is distance (Jm). After Greaves, W. S. (1991b). The orientation of the force of the jaw muscles and the length of the mandible in mammals. *Zoological Journal of the Linnean Society*, **102**, 367–374. (b) When the distance from the joint to the third molar (m) is held constant, the jaw with an anterior vector is longer than the jaw with a posterior vector.

Representative muscle resultants with anterior (A) and posterior (P) inclinations can be chosen where, as in the earlier analyses, the distance from the jaw joint is three units in each case and the muscle force is equal to one unit. Thus the moments of both of these muscle forces are the same ($3 \times 1 = 3$). Two vectors are shown in Figure 3.3a for ease of comparison. The intersection of the muscle vector with the long limb of the L (at **a** and **p**) determines the rear of each tooth row. Recall that all the teeth are always in front of the vector. The intersections of the line along the tooth row with the lines that are parallel to, and seven units from, each vector define the anterior end of each tooth row (**a′** and **p′**). Measurements are taken *perpendicular* to the muscle vectors, and these distances are projected onto the tooth row. It is the perpendicular distances of 3 and 7 that give jaws a length of ten units.

Therefore, in Figure 3.3a neither tooth row is seven units long when measured *along* the row because the vectors themselves are not perpendicular to the row. This situation is typically the case in mammals.

Note that the absolute length of a tooth row varies with the inclination of its vector. However, by adjusting the inclinations of an anterior and a posterior vector, tooth rows of exactly the same length can be achieved (**aa′** = **pp′**). This maneuver facilitates a comparison between the two because relative bite forces are then the same at equivalent positions along each tooth row.

The distances from the joint (J) to the intersection of the vector with the tooth row (at **a** or **p**) differ in these two cases (dashed lines). Line **Jp** is much longer than line **Ja**. These two distances are functionally important because they indicate the length of the bony connection between the jaw joint and the tooth row. A longer distance naturally requires a longer bar or arches of bone in both the skull and lower jaw.

Short beams, whether made of steel, wood, bone, or some other material, are in one sense better than longer beams. When resisting given forces with limited distortion, more material is needed to construct a longer beam, not only because it is longer, but also because additional material is required to provide it with the same rigidity as a short beam doing an equivalent task.

Bone, like many other tissues in the body, is metabolically expensive. It must be built or constructed in the first place and is then constantly remodeled as bone tissue is removed and added. Therefore, bones are generally considered to be as short, or as economically built, as possible. Thus, one would anticipate that the distance between the joint and the tooth row would be as short as possible. This seems to be especially true in the example in Figure 3.3a. The anterior and posterior vectors have the same moments and produce exactly the same bite forces at equivalent positions all along their respective tooth rows because these rows, in this particular case, are equal in length.

Distance **Ja** is shorter than distance **Jp** when the jaw joint, as in most mammals, lies above the level of the tooth rows. Therefore, less bone tissue, and thus less metabolic effort, will be required to span distance **Ja**. All of this suggests that, when possible, muscle force vectors that are anteriorly inclined will be selected for over posteriorly inclined vectors, because less bone tissue is required to join the jaw joint to the tooth row in these cases.

Of course, reaction forces at both the teeth and the joint are inclined in different directions in these two cases. These inclinations could be important. For example, some researchers have suggested that posteriorly oriented tooth forces might be expected in animals such as carnivores. In some cases struggling prey animals may exert anterior forces on the lower jaw that may have to be resisted by a large posterior force provided by a large, posteriorly inclined, temporalis muscle.

The distance from the jaw joint to the third molar is shorter when the muscle vector is anteriorly inclined (Fig. 3.3a). However, there is a limit to how short this distance can be. It cannot be shorter than three units when a ten-unit-long jaw is considered, because the muscle force is 30% of the way along the jaw. The distance

Jm in Figure 3.3a represents this minimum distance from the joint to the third molar. The vector (not shown) that is associated with this (perpendicular) distance from the jaw joint is anteriorly inclined. Distances from J to points along the tooth row that are posterior to **m** are less than three units from the joint. However, such points do not correspond to real animals, where the distance to any resultant muscle vector is three units. Note that the vector associated with the shortest distance to the joint, the horizontal line along the tooth row, and the line from the joint to the third molar (**Jm**), all intersect at the *same* point. Recall that the vector that intersects the horizontal at **m** is not shown.

[Before proceeding with the analysis, note that the relationships just described can be examined from a different point of view. In the analysis above, the distance from the jaw joint to the muscle vector was held constant. The distance from the joint to the third molar then varied with the inclination of the vector. Suppose the distance from the *joint to the molar* is held constant (Fig. 3.3b). In this case, all the vectors intersect the tooth row at the same point (**m**). The perpendicular distance from the *jaw joint to the vector* then varies with the inclination of the vector. The projected length of the jaw also varies because the distance from the joint to the vector is 30% of projected jaw length. Thus, when the joint is above the level of the teeth, anteriorly inclined vectors (e.g., A) will be associated with long tooth rows (e.g., **ma**′), whereas posterior vectors (e.g., P) will be paired with short rows (e.g., **mp**′). (If a joint below the tooth row is assumed, anterior vectors will have a short row and posterior vectors will have a long row.) In either case, real skulls and jaws of approximately the same size are required for a proper comparison.]

Viewing the jaw mechanism from the first point of view suggests that given the same jaw length, anteriorly inclined vectors are better because less bone is required to join the jaw joint to the tooth row. Yet, the second viewpoint suggests that a posterior vector requires less bone tissue if the distance from the joint to the molar is the same and the lengths of the tooth rows are different. The entire jaw is shorter when the vector is posteriorly inclined (Fig. 3.3b); an anterior vector defines a much longer jaw. Accepting that the length of the tooth row is the most important consideration and the distance from the joint to the tooth row is of secondary importance implies that the former approach is more appropriate, and thus an anteriorly inclined muscle vector should be selected for. Nevertheless, this relationship between the inclination of the vector and the distance from the joint to the molar is a critical issue for the analysis that will be examined in detail in section 3.5.1.

To say that a vector is anteriorly or posteriorly inclined simply means that it forms some angle with the line that passes, approximately, along the tooth row (Fig. 3.4a). Suppose that the jaw joint (J) is positioned at the end of a line that is three units long that begins at the intersection of the vector and the tooth row. The distance from the joint to the third molar is then as short as possible. Distance is a perpendicular measure, so this line forms an angle of 90° with the vector. Thus this line represents a distance of three units from the vector to the joint. The angle (α) between the horizontal line representing the tooth row and a vector must be less than

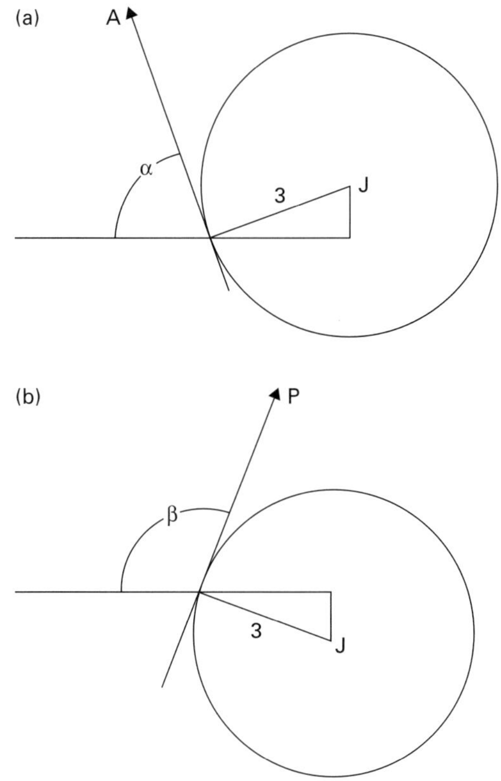

Figure 3.4. Anterior (A) and posterior (P) vectors intersecting horizontal tooth rows. The anterior angle between the vector and the tooth row is smaller than 90° in (a) and greater than 90° in (b). The circles have radii of three units. The joint lies above the level of the tooth row when the vector is anteriorly inclined. The joint lies below the level of the tooth row when the vector is posteriorly inclined.

90° if the vector is to be anteriorly inclined. The sum of the two angles (α + 90°) is therefore less than 180°, and the jaw joint lies *above* the tooth row.

The angle (β) between the tooth row and a vector must be greater than 90° if the vector is to be posteriorly inclined (Fig. 3.4b). The line to the joint again begins at the intersection of the vector and the tooth row, and the distance from the molar to the joint is three units. The angle between the vector and the line to the joint is again 90°. In this case the sum of these two angles (β + 90°) is greater than 180°, and the joint is located *below* the tooth row.

[Another way of examining this idea is to consider two lines representing an anterior and a posterior vector. Both vectors intersect the horizontal line representing the tooth row at the same point. Imagine two additional lines that are parallel to, and at a distance of 30% of jaw length from, each vector. Then consider a circle centered at the intersection of the vectors with the tooth row and tangent to each of the two additional lines. The radius of the circle indicates the shortest distance to these two parallel lines. This procedure clearly demonstrates that an anterior vector has a joint *above* the level of the tooth row and a posterior vector has a joint *below* this line.]

Thus, the distance from the jaw joint to the third molar can be minimized (i.e. set to three units) regardless of the inclination of the vector. However, an anterior vector requires a joint that is *above* the tooth row, whereas a posterior vector requires a joint that is *below* the tooth row. Almost all mammals have a joint that is some distance *above* the level of the tooth row. Clearly then, in mammalian groups with a posteriorly inclined muscle vector (the majority), assuming tooth row length is a primary measure, these analyses suggest that the distance from the jaw joint to the third molar is not as short as it could be (cf. Figs 3.3 and 3.4).

The analysis to this point indicates that important distances, such as the distance between the jaw joint and the third molar, seem *not* to be as short as possible because many mammalian groups have jaws with a resultant muscle force vector that is *posteriorly* inclined and the distance from the joint to the molar is longer than the theoretical minimum. Animals with *anteriorly* inclined resultant muscle forces can theoretically approach this minimum distance. Apparently, more bone tissue is needed to construct the skull and lower jaw than would be expected when assuming a minimum amount of bone tissue.

3.4.2 Minimizing a pair of distances

In order to continue the analysis, a number of different but theoretically possible locations for the jaw joint can now be considered. The posterior unit vector (P), with a line-of-action MT, is inclined toward the rear in Figure 3.5. The perpendicular distance from the vector to each of the *hypothetical* joints (J1–J6) is always 30% of jaw length. Therefore, all of these hypothetical joints will lie on a single line that is parallel to, and 30% of jaw length from, the line-of-action of the vector of muscle force (P). The joint can be positioned above, below, or on the same level as the tooth row. In each case its (perpendicular) distance to the vector is the same: three units.

Yet, the distance of the joint to the third molar (M), which represents the bony connection between the joint and the tooth row, is different for each joint location. The solid line from M to the joint at J6 represents the shortest possible distance between the two. As mentioned earlier, this joint (J6) lies below the level of the tooth row because the vector in this example is posteriorly inclined. The distance from each hypothetical joint to the molar (M) is indicated in solid line. In all cases, except for that at joint J6, this distance is longer than the minimum. Again, because the jaw joint is located above the level of the tooth row in almost all mammals, the distance between the joint and the molar is not minimized in the large number of actual cases in which the vector is posteriorly inclined.

Nevertheless, the assumption that apparently functionally important struts or plates of bone should be as short as possible, so as to minimize the amount of bone tissue, still seems plausible. However, the length of this particular distance from a jaw joint to the molar is quite obviously not as short as possible in the large number of mammals that have posteriorly inclined muscle vectors. This suggests that

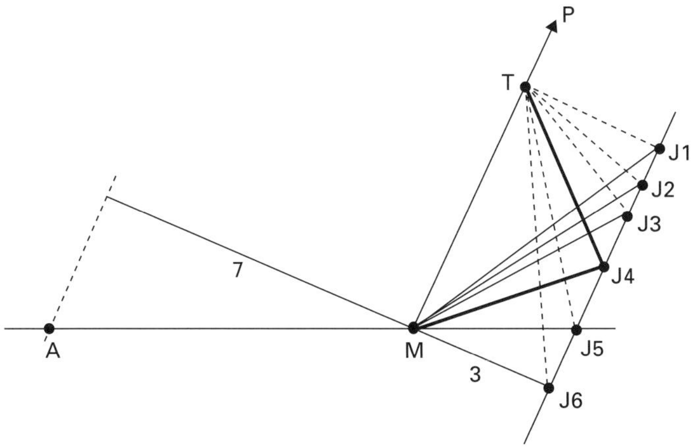

Figure 3.5. A diagrammatic jaw viewed from the side with six hypothetical jaw joint locations, demonstrating that the combined distance (MJT) from the third molar (M) to the joint (J) to the point T is a minimum length when the jaw joint is located at J4. After Greaves, W. S. (1998). The relative positions of the jaw joint and the tooth row in mammals. *Canadian Journal of Zoology*, **76**, 1203–1208.

perhaps minimizing a single distance in the skull is not sufficient. There are other important distances in the skull and jaw that also may be expected to be a minimum length. Thus, examining two functionally important distances *at the same time* may be more appropriate.

In Figure 3.5 the unit vector intersects the tooth row just behind the third molar (M). A single point (T) located on the line-of-action of this vector of muscle force can be chosen for analysis. Such a point would lie somewhere within the fossa on the side of the skull, where the temporalis muscle attaches. As the muscle vector pulls down on the skull as well as up on the lower jaw, the distance from the joint J (1–6) to point T is also a functionally important span. This region of the side of the skull resists both the downward pull of one of the jaw muscles and the upward push of the condyle of the lower jaw. Typically this part of the skull in mammals is well suited to resist compressive forces of this sort because it is usually made up of two curved plates that form the inner and outer tables of the braincase and thus constitute the sidewall of the posterior skull.

The dashed lines in Figure 3.5 that extend from each hypothetical joint to point T represent the curved plates of bone that are compressed by the action of the temporalis muscle. These lines represent distances that perhaps should also be as short as possible. Shorter distances reduce the amount of bone tissue as well as the metabolic cost associated with its upkeep. Needless to say, a sufficient volume within the skull for the brain is assumed. Therefore, this distance cannot become too short. (Shortening this distance (JT) might reduce the area on the side of the braincase that is required for muscle attachment. The development of bony crests or ridges (i.e., the sagittal and nuchal crests in some carnivores) is one way to compensate for reduction in area in this region; muscle attachment area is increased, but in a less metabolically expensive way than by increasing the size of the *entire* braincase.)

In Figure 3.5, for the series of different hypothetical joint positions, these two distances (JMs and JTs) are indicated. The distances JM from each joint to the molar are indicated in solid line and distances JT from each joint to point T on the vector (P) are indicated in dashed line.

Examination of this figure demonstrates that only one of the distances in each of these distance pairs can be as short as possible. The high joint at J1 is as close as possible to point T but is far from the molar at M. As lower and lower joint locations are considered, the distance to the molar decreases, but at the same time the distance to point T increases; when one distance decreases the other increases. Therefore, both of these distances cannot be minimized at the same time. In the case of the highest joint (J1), the distance to T is short whereas the distance to M is long. The opposite is true for the joint (J6) just below the level of the tooth row.

An obvious solution to this difficulty is to minimize not one or the other of these distances, but rather both together so that their *sum* is as short as possible. Inspection (and analytic geometry) indicates that the sum of the two distances is minimized when the jaw joint is located such that the two individual distances (JT and JM) are equal. In the example in Figure 3.5, the equal distances are represented in heavy line. A joint in position J4 therefore minimizes the *sum* of the two distances.

Note that if a different orientation of the vector is considered, or if a different point T is chosen, the location for the jaw joint that minimizes the sum also changes. Equal distances are required regardless of the location of point T. Nevertheless, for all the inclinations that a posterior vector is likely to have in a real mammal and for all reasonable locations for point T, the jaw joint is almost always positioned *above* the level of the tooth row.

This might appear to be the most noteworthy result of this particular analysis because it provides a reason (a minimized distance) for a joint being above the level of the tooth row in mammals. Another analysis below will suggest the same thing (see section 3.4.4, below). However, so far, only two functionally significant distances have been considered. This idea about minimizing the amount of bone tissue required to form struts or plates in the skull and jaw suggests that other pairs of distances can be found, especially in a complicated structure such as the skull. At the same time, however, if this system is to be modeled, and the model is to be tractable, a fairly small number of such distance pairs would be welcome.

The analysis so far implies that the direct distances from the joint to the molar and from the joint to some point T are equal. In a real mammal the locations of the joint and the ends of the tooth row are known. Moreover, as described earlier, the line-of-action of the resultant vector of jaw muscle force can be estimated by rotating a line around the point M at the molar until the perpendicular distances, from this line to the joint and to the anterior end of the jaw, are in the ratio of 3:7. Once the line-of-action of the vector is determined, an arc centered on the joint, with a radius equal to the distance from the joint to the molar, can be drawn. The intersection of this arc with the line-of-action of the unit vector is point T.

3.4.3 Using four points to model the jaw mechanism

Given that there is one compound structural distance MJT (molar to joint to point T) that can be minimized in a skull, then presumably there are other such distances in this complicated structure. One approach to finding them is to first define the jaw mechanism as simply as possible. Then, the simple model can be examined carefully to see whether other such distances suggest themselves.

Previous analyses have concentrated on a few small regions that are considered to be points for ease of analysis. In a skull, points are defined at the jaw joint and both ends of the tooth row. The points M and A therefore define the tooth row, and JM and JA locate the joint (J) relative to the row. As indicated earlier, the resultant unit vector of jaw muscle force divides a jaw in a 3:7 ratio. For any given three points (J, M, and A) that do not lie on the same line, there is only one vector with a line-of-action that divides the jaw in this way. Using only these three points and the line-of-action of a vector, relative bite forces can then be calculated all along a tooth row.

As mentioned earlier, because we know the line-of-action, the point T can be located on the vector by setting JT equal to JM. The distance from M to J to T is minimized because MJ is equal to JT.

All the distances between these four points seem to be functionally significant. JM connects the joint to the tooth row (MA), JA estimates the entire length of the upper jaw, AT resists the rotation of the anterior, upper jaw upward toward the braincase during biting, and TJ resists compressive forces produced by the jaw muscles. The distance from T to M varies with the inclination of the vector and also resists compressive forces. The tooth row (AM) is specific to a given animal and is presumably not to be minimized. Note that point T does not define the length of the vector and therefore does not define its magnitude; the vector is a unit vector.

These four points (A, M, J, and T) define the basic jaw mechanism. Relative bite forces can be calculated if these four points are known. Some of these distances perhaps should be as short as possible. For example, if the locations of points J, M, and A are determined on a real animal, we readily see that AT appears to be equal to AJ (Fig. 3.6). Thus, the sum of these two distances (compound distance JAT) might also be minimized.

With this idea in mind, a second way of locating point T is by drawing an arc centered on A with radius AJ (cf. Fig. 3.6). The intersection of this arc and the line-of-action is point T. Moreover, line AJ is equal to line AT, and thus the compound distance JAT is minimized. Thus two arcs, one centered on J with radius JM and the other centered on A with radius AJ both define the same point T. That is, both arcs intersect at the same point on the line-of-action.

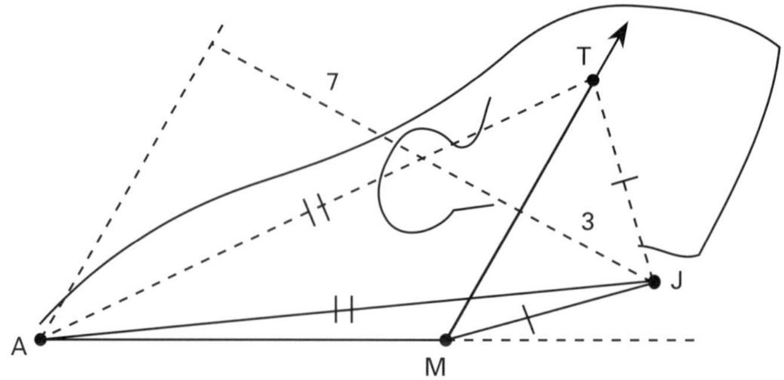

Figure 3.6. A diagram indicating the equal distance pairs (JM and JT; AJ and AT) used to define point T. In this case the jaw joint is fairly low relative to the tooth row. The outline of a diagrammatic skull is included for orientation. After Greaves, W. S. (2004). Estimating the line of action of posteriorly inclined resultant jaw muscle forces in mammals using a model that minimizes functionally important distances in the skull. In: *Shaping Primate Evolution*. ed. F. Anapol, R. Z. German, and N. G. Jablonski. Cambridge: Cambridge University Press. pp. 334–350.

Line TM is therefore *coincident* (see section 3.4.5 below) with the line-of-action of the unit vector, as determined by measuring the distances between the vector, and the joint (three units) and the anterior end of the jaw (seven units). Note that the two arcs also intersect at a second point located below line JM (cf. Fig. 3.6). This point does not define a structural entity, as it lies outside the confines of the skull. Obviously, two arcs can be drawn from each of the three points A, M, and J. Most of the intersections of these arcs lie outside the skull and thus do not define meaningful T-like points.

[Yet, there is another meaningful point. Two arcs with a radius of length JM can be constructed. One arc is centered on J and the second arc is centered on M. The intersection of these two arcs is located below, and just anterior to, point T. This point, together with points M and J, forms an *equilateral* triangle at the deep lateral side of the skull. These three minimum distances potentially brace the front and rear halves of the skull at the weak orbital region.]

3.4.4 The spatial configurations of the four points

The resultant muscle unit vector lying along MT divides the distance from the joint (J) to the end of the jaw (A) in the ratio of 3:7 when both of these distances are perpendicular to the vector, as described earlier (Figs 3.6 and 3.7). If the muscle vector is posteriorly inclined, then the angle it forms with the tooth row (angle TMA) is more than 90°. Finally, the jaw joint is almost always located above the level of the tooth row. Given the above, angle JMA will typically be no larger than approximately 178° (it has to be somewhat less than 180° if the joint is to be above the tooth row). The following graphical analysis of the model determines the range of

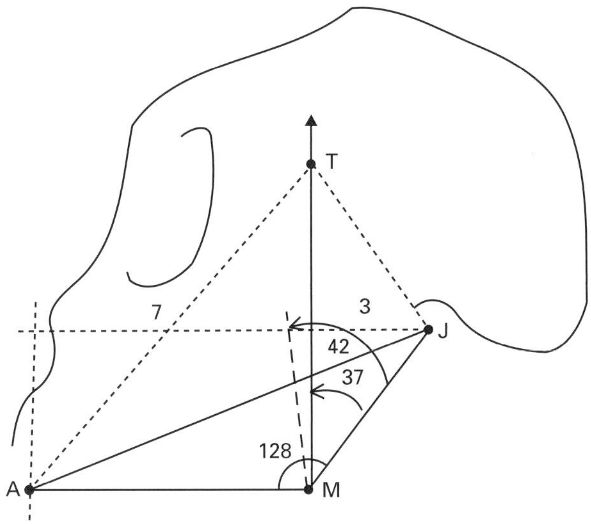

Figure 3.7. A diagram indicating the equal distance pairs (JM and JT; AJ and AT) used to define point T. In this case the jaw joint is high above the tooth row. In general, the line-of-action of the vector (arrow) lies about 5° to either side of a line that forms an angle of 42° with the line from the joint (J) to the molar (M). In this case the vector forms an angle of 37° with line JM. The outline of a diagrammatic skull is included for orientation. After Greaves, W. S. (2004). Estimating the line of action of posteriorly inclined resultant jaw muscle forces in mammals using a model that minimizes functionally important distances in the skull. In: *Shaping Primate Evolution*. ed. F. Anapol, R. Z. German, and N. G. Jablonski. Cambridge: Cambridge University Press. pp. 334–350.

possible configurations that can be taken by the four points (J, M, A, and T) when viewed from the side.

To determine this range, three parallel lines are constructed. The intervening distances between these lines are in the ratio of 3:7. A point is then chosen on the line in the center to represent point M; point J must be located somewhere on the line that is three units from the center line; point A must be somewhere on the remaining line that is seven units away; and point T is located above point M on the center line. The range of appropriate configurations of the four points can be determined by finding a sample of locations for J, M, A, and T on their respective lines. However, the following must be true: (1) distance JT = JM and distance AT = AJ (to minimize the distances); (2) angle TMA is greater than 90° (to insure a posterior inclination for the muscle resultant); and (3) angle JMA is less than 180° (to insure that the joint lies above the level of the tooth row).

One can start with a point M on the line in the center and choose a point J on the line that is three units away. Point T, on the centerline, is thus determined because JM = JT. As AJ must equal AT, plane geometry indicates that the perpendicular bisector of the line from J to T intersects the third line at point A. The 3:7 ratio holds because all the points lie on their respective lines, where the centerline divides the space between the other two lines in the ratio of 3:7. If point J lies above the

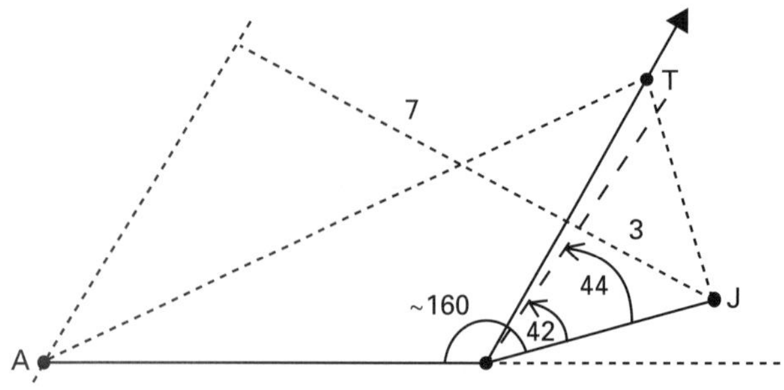

Figure 3.8. Equal distances JM and JT and distances AJ and AT are used to define the location of point T. The line-of-action forms an angle of 44° with line JM. A diagrammatic skull outline is omitted. After, Greaves, W. S. (2004). Estimating the line of action of posteriorly inclined resultant jaw muscle forces in mammals using a model that minimizes functionally important distances in the skull. In: *Shaping Primate Evolution*. ed. F. Anapol, R. Z. German, and N. G. Jablonski. Cambridge: Cambridge University Press. pp. 334–350.

extension of AM and at the same time angle TMA is greater than 90°, the joint lies above the tooth row and the vector is posteriorly inclined.

Of all the many possibilities, only cases that meet these requirements lie within the acceptable range. A sample of appropriate configurations that fall within the acceptable range can be determined by trial and error.

After determining the range of appropriate configurations, a number of relationships can be noted. Near one end of the range, the muscle vector is almost normal (90°) to the tooth row because it is just barely posteriorly inclined. The jaw joint is very high above the tooth row in this case (Fig. 3.7). At the other end of the range, the vector is markedly inclined toward the rear. Here the joint is only slightly above the level of the tooth row (cf. Figs 3.6, 3.7, and 3.8).

Consider a line that makes an angle of 42° with line JM (long dashes in Figures 3.7 and 3.8). The vector is restricted to about 5° either side of this line. Thus, the vector makes an angle of approximately 37° with line JM when the joint is high (Fig. 3.7) and approaches 48° with line JM when the joint is low. In the case diagrammed in Figure 3.8, the joint is a short distance above the level of the tooth row, and thus angle JMT is approximately 44°. Angle JMA varies from a low of about 128° to a high of about 178° and is easily measured on mammalian skulls.

At the upper end of the range, this angle is just short of 180° because the jaw joint is *observed* to lie above the tooth row in most mammals, so this angle is not derived (Fig. 3.8). At the lower end of the range this angle is approximately 128° because it is the sum of 37° and an angle just larger than 90° (Fig. 3.7). Because the latter angle is slightly larger than 90° the vector will have a posterior inclination. This minimum angle of 128° is approximately the minimum that is found in a large sample of primates and pigs.

[A large tendinous sheet extends from the dorsal and posterior edges of the coronoid process. This sheet increases the effective size of the bony process. Both the hard and soft parts of this process provide a surface for the attachment of most of the fibers of the large temporalis muscle. By examining the bony coronoid process, one can get a rough idea of the shape of this sheet and therefore some idea of the direction in which this major muscle is pulling. The angle the anterior edge of the coronoid makes with a line from J to M has *not* been studied in a large number of animals. Nevertheless, a cursory examination suggests that this anterior edge forms an angle that approaches 42° with line JM. This angle therefore gives a *very rough* estimate of the line-of-action (MT) of the muscle force. This appears to be the case even to the extent that the angle between the edge of the coronoid and line JM is somewhat smaller than 42° in animals with high joints and slightly larger than this in those animals with low joints.]

When the joint is very high, the vector is almost perpendicular to the tooth row. The component of force normal to the tooth row is very large, whereas the posterior component is very small. For one unit of muscle force, the normal force at a tooth near the vector's intersection with the tooth row is almost one unit. As the jaw joint gets closer to the level of the tooth row, the rearward inclination of the vector approaches 50°. Thus, the posterior component of muscle force gets very large, although it never gets as large as the vertical component that is normal to the tooth row. About 1.3 units of muscle force are needed to produce one unit of force normal to the tooth row. If the inclination of the vector increased any farther to the rear the joint would be below the tooth row and the posterior component of muscle force would be larger than the normal force. Given the importance of bite force, such very large posterior components are unlikely. This circumstance implies that the jaw joint will rarely be located below the level of the tooth row.

In summary, the height of the joint above the level of the upper tooth row varies with the inclination of the resultant vector. When the vector has an appreciable posterior inclination the joint is very low. An almost vertical vector is associated with a very high joint.

Only certain spatial configurations of the glenoid cavity (J) and the two ends of the upper tooth row (A and M) are found within the range of possibilities, as defined by the model described above. The three points in each configuration naturally define a triangle AMJ. At one extreme, where the glenoid is high, the angles of such a triangle are approximately 128, 30, and 22° (Fig. 3.7). At the other extreme, where the glenoid is low, the triangle could almost collapse to a line and angle M would approach 180°, whereas angles A and J would approach zero. An infinite series of triangles are possible that will span the entire range of possible skull sizes. Beginning with a very low glenoid, angle M would start at slightly less than 180 and decrease to 128° whereas angles A and J would increase from near zero to about 22 and 30°, respectively. As these angles increase, the height of the glenoid, or the perpendicular distance from the level of the tooth row up to the joint, would also increase.

Triangles with angles varying from 128° to near 180°, as just described, represent only a small subset of all mammals. To represent most mammalian jaws, larger and smaller triangles, at every step, are also required. At each increment these additional triangles are geometrically similar, with equal equivalent angles but with longer or shorter sides. Thus, a unique triangle can be drawn that will represent a particular animal of any size and appropriate glenoid height provided that the vector is posteriorly inclined.

3.4.5 Point T lies on the vector

Notice that beginning with any group of three points that are not colinear, the following *two* lines can be constructed. The first is a line that begins at the middle point. The ratio of the perpendicular distances from this line to the two remaining points is 3:7. This is the line that estimated the line-of-action of the resultant vector in a number of the analyses above. The second line also begins at the middle point and extends to the intersection of two arcs. The two outer points serve as the centers of these arcs. The intersection of the arcs is point T in previous analyses. When all triplets of three points not on the same line are considered, sometimes these two constructed lines are coincident; often they are not. If the lines are not coincident, one can imagine either: (1) the two lines are at some angle to each other; or (2) the two arcs intersect the line-of-action at two different points.

When mammalian jaw mechanisms are examined, the lines determined by these two methods are coincident (and, significantly, agree with the line-of-action determined from dissection of the jaw muscles). Thus, point T lies on the line-of-action of the resultant vector. This seems to be the usual case (cf. Figs 3.7 and 3.8). Obviously not all mammals have been examined so far, but a sample representing a number of mammalian orders strongly suggests that this is the general situation.

Accepting that minimized distance pairs are important parameters for the construction of the mammalian jaw mechanism, these two lines are expected to be coincident. This relationship is evidence that minimizing distances, suggesting a minimum amount of bone tissue, is a reasonable approach.

Given the four points that define the jaw mechanism (A, M, J, and T), the length of the tooth row (AM) is peculiar to the animal being examined and the length of the line from M to T varies with the inclination of the vector (cf. Figs 3.7 and 3.8). The remaining four of the possible six lines defined by these points (AJ and AT; JM and JT) are minimum distance pairs.

In Figure 3.7, two distances (MJ and AJ) locate the posterior and anterior ends of the tooth row (MA) relative to the joint (J). A previous analysis determined that given point T, on the line-of-action of the vector, the joint should be located such that the distance from M to J is equal to the distance from J to T (Fig. 3.5). Two equal distances extend from the *anterior* end of the tooth row at A to the joint at J and to point T on the line-of-action of the vector (Figs 3.7 and 3.8). Extending the analysis in Figure 3.5, when points J4 and T are known, allows the position of the anterior end of the tooth row (A) to be determined. The perpendicular bisector of line JT intersects a line parallel to, and 70% of jaw length from, the vector MT (cf. Figs 3.5 and 3.8). Of course, the resulting figure must conform to the spatial configurations of the four points as previously described.

Note how the joint and the two ends of the tooth row are configured as indicated in Figures 3.7 and 3.8. The line along the tooth row (from M to A) and lines from the joint (J) to both M and A form a triangle in each case. In the case in which the joint is high and the muscle vector is almost perpendicular to the tooth row (Fig. 3.7), the triangle formed by these three points is obtuse, with the angle at M approximately 128°. In a case near the other end of the range, where the joint is low and the vector has a strong backward inclination (Fig. 3.8), the triangle is also obtuse, but the angle at M is large and approaches 180°.

Each side of these triangles can be compared with their perimeters (the sum of all three sides). The triangles described by the model vary incrementally in shape because they form a limited series. The side JA in the models accounts for approximately half of the perimeter (0.48 in the case of the high joint and 0.50 when the joint is fairly low). There is slightly more variability in the length of the tooth row (MA), which varies from 0.31 to 0.35 in the cases of high and low joints, respectively. The distance from the joint to the molar (JM) is more variable because this measure is heavily influenced by the height of the jaw joint (0.22 to 0.15 in high and low joints, respectively). Triangles outside the range of those described by the model have sides that depart more and more from the ratios noted above.

To get some idea of how the model corresponds to real animals we can examine measurements of primate skulls in Spencer's 1995 dissertation (see the References section). Lines approximating JM, JA, and MA can be calculated from his measurements, taken from around 200 skulls representing 10 different genera and 12 species of primates. Spencer's measurements do not allow the calculation of the exact distances required by the models (the calculated JM is slightly too long, MA is slightly too short, and JA is about right). Yet these calculated distances are reasonably close to what is needed and are worthwhile because of the large number of measurements involved.

Using these measurements from primates, the distance from the joint to the anterior end of the jaw (JA) divided by the sum of JM, JA, and MA (the perimeter of the triangle) equals either 0.49 or 0.50 in all the species examined, except in one case, where this quotient is 0.48 in both males and females. This is essentially the same range as in the model (0.48 to 0.50). The tooth row (MA) as a fraction of the perimeter varied from 0.26 to 0.32 compared with 0.31 to 0.35 in the model. Given that the calculated distances are close to, but not exactly, what is required, this fraction is expected to be slightly lower. The more variable distance JM when divided by the perimeter varied from 0.19 to 0.24, except for the males in one genus (0.26). As noted these calculated quotients are expected to be a little larger than the figures from the model (0.15 to 0.22). These results taken together suggest that the range of triangles derived from a large number of primates and from the model are much the same, keeping in mind that Spencer's measurements allow the calculation of distances that are only approximately that required for a strict test.

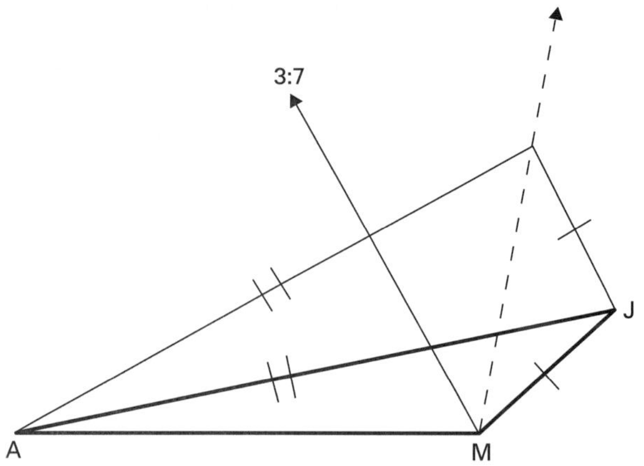

Figure 3.9. A diagram indicating the joint (J), the third molar (M), and the anterior end of the jaw (A). The left-hand arrow indicates a constructed anterior vector that divides the jaw in a 3:7 ratio. The right-hand arrow indicates a posterior vector defined by the intersection of an arc centered on J with radius JM with an arc centered on A with radius AJ. After Greaves, W. S. (2004). Estimating the line of action of posteriorly inclined resultant jaw muscle forces in mammals using a model that minimizes functionally important distances in the skull. In: *Shaping Primate Evolution*. ed. F. Anapol, R. Z. German, and N. G. Jablonski. Cambridge: Cambridge University Press. pp. 334–350.

3.5 Anteriorly inclined vectors

Many mammals, such as antelopes and rodents, have resultant vectors of jaw muscle force that are anteriorly, rather than posteriorly, inclined relative to the tooth row. That is, the angle anterior to the vector, between it and the tooth row, is less than 90°. In most of these animals the temporalis muscle is smaller than the masseter and pterygoid taken together. These larger, anteriorly inclined muscles tilt the resultant vector of muscle force anteriorly.

When a large masseter/pterygoid complex is present and points J, M, and A are known, the appropriate *anteriorly* inclined line-of-action of the vector can be determined relatively easily as described earlier. Simply rotating a line around point M until a 3:7 ratio is attained between the distances from the line to the joint, and to the anterior end of the jaw, gives a good estimate of the line-of-action of the vector. However, if the determination of the anteriorly inclined line-of-action is attempted using intersecting arcs (with radii JM and AJ), in the manner described previously (see sections 3.4.3 and 3.4.4), a *posterior* vector is defined that does *not* divide the jaw in a 3:7 ratio (Fig. 3.9); the two lines are not coincident.

From another point of view, these arcs intersect the already determined *anterior* vector at two different points. Thus, this procedure (intersecting arcs) used to define a posterior vector fails to define an appropriate anterior vector.

3.5.1 A relationship between an anterior vector and a long diastema

The above indicated that using intersecting arcs to determine the inclination of anterior resultant vectors of jaw muscle force, in a way that is analogous to that used for posterior vectors, is apparently unworkable. The evolution of anterior vectors, and even why there are vectors that are anteriorly inclined, can perhaps benefit from a more detailed examination.

A dominant temporalis, and thus a posteriorly inclined muscle vector, is present in most orders of mammals. As noted earlier (in sections 3.4.3 and 3.4.4), less bone is required to construct the jaws in these animals. Some have suggested that in carnivores the muscle vector's posterior component is important in resisting anterior forces when an animal contends with struggling prey.

An anterior muscle vector is found in only a few orders of mammals, although these are large groups with many species. An anterior vector is present because the masseter and pterygoid are large, and these muscles have anterior inclinations. The temporalis, with its posterior inclination, tends to be smaller in these few groups.

In the modern forms, within these groups, in addition to an anterior vector, the anterior parts of the jaws appear to have lengthened when compared with the jaws of primitive, or early, members of these same groups. In modern antelopes and their kin, a lengthened snout may be related to improving the ability to select particular types of vegetation. In addition, as indicated above, higher bite forces on some premolars because these teeth are included in a longer Region II also seems to be important. Rodents and rabbits also have long anterior jaw regions, but this is likely to be due mainly to the development of extremely long and ever-growing incisor teeth.

Antelopes and similar animals also have a toothless region within the tooth row. This diastema is extremely long and is clearly not the result of the loss of a few teeth. Rather, it seems to be related to a longer snout coupled with teeth that, for the most part, maintained their size so that they no longer fill the lengthened jaw. A similar situation is also found in a few species in other mammalian orders (e.g., the horse).

The difficulty that arises when the anterior part of the jaw lengthens is that this could potentially disturb the 3:7 ratio. Thus, the largest sum of bite forces would not be realized. A concomitant lengthening of the posterior part of the jaw might be expected in order to maintain the ratio. That maneuver would move the jaw joint to the rear, restoring the ratio. Alternatively, the vector could move anteriorly in order to position the muscle force 30% of the way along the longer jaw. Because the teeth are located in front of the vector, the tooth row would also have to be adjusted. Either of these changes would entail a very serious disruption of the bony structure of the skull and jaw. A third alternative requires no further change in the joint or the teeth or any of the skeletal parts of the skull and jaw. All that is required is a relatively minor change in the *masses* of the three great muscles that close the jaw. That is, a muscle or muscles would get a little larger or a little smaller.

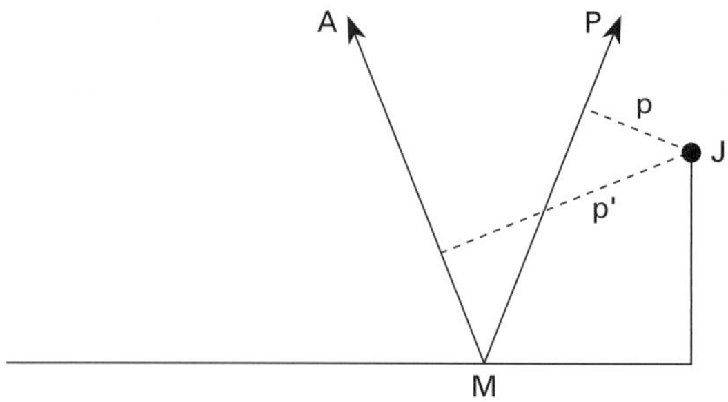

Figure 3.10. A diagram to indicate that when the jaw joint is positioned above the level of the tooth row the moment arm (p) of a posterior vector (P) is shorter than that (p′) of an anterior vector (A).

Recall that the 3:7 ratio refers to the *perpendicular* distances, from the line-of-action of the resultant vector of jaw muscle force to the jaw joint (three units of distance) and to the anterior end of the tooth row (seven units of distance). The three-unit-long, posterior region is toothless, whereas the teeth and the diastema are located in the longer, anterior part of the jaw. The direct distance from the joint to the rear of the tooth row and the actual length of the tooth row itself are not the relevant measures; only the previously mentioned perpendicular distances (projections) are important in this analysis.

Imagine that a change in the masses of the jaw muscles (say, a decrease in the temporalis and an increase in the masseter and pterygoid) shifts the inclination of the muscle vector more anteriorly (from vector P to vector A in Figure 3.10). This figure clearly demonstrates that such a shift increases the *perpendicular* distance (**p**) that extends from the joint to the vector. Thus, distance **p** increases to **p**′. Of course, these distances are the shorter (**p**) and longer (**p**′) moment arms of the respective vectors. Naturally, the perpendicular distance from the vector to the anterior end of the jaw also changes. Note that no bony parts of the skull or jaw, beyond the initial jaw lengthening, have changed in any way. Only the masses of the jaw muscles are different (smaller temporalis, larger masseter and pterygoid, or a little of both).

The previous paragraph suggests that the lengthening of the anterior region of the jaw in a few groups of mammals, for whatever reason, is not a very serious problem after all. Such an increase will not disturb the 3:7 ratio as long as there are relatively minor concomitant changes in the muscle masses that rotate the resultant vector of muscle force anteriorly by an *appropriate* amount. Recall that estimating the location of a vector that divides a jaw in a 3:7 ratio only requires that it be rotated around a point at the third molar until this ratio is achieved. Thus, rotating the vector by changing muscle masses maintains the 3:7 ratio in a remarkably simple fashion.

An obvious evolutionary scenario is as follows. When the anterior portion of the jaw elongates for some reason, a diastema forms because the teeth have remained much the same size and therefore no longer fill the longer anterior portion. A simultaneous change in muscle masses takes place that shifts the inclination of the resultant vector anteriorly. This change in orientation adjusts the perpendicular distances, from the vector to the joint and from the vector to the anterior end of the jaw, just enough to maintain the 3:7 ratio. That is, the vector rotates anteriorly until the 3:7 ratio is achieved between the distances to the joint and the anterior end of the jaw as previously described.

In effect, the increase in anterior jaw length, and its perpendicular distance, is balanced by a moment arm increase (also perpendicular) due solely to vector orientation, resulting in the maintenance of the 3:7 ratio. The orientation was produced by changes in muscle masses rather than any bony changes in the skull and jaw beyond the original jaw lengthening. A dominant masseter/pterygoid complex produces an anteriorly inclined vector.

Very long incisor teeth generally accompany an anterior vector in rodents and rabbits. In antelopes and their kin, as well as horses, an excessively long diastema is present. Moreover, a postorbital bar develops in some of these latter animals. Such bars resist the twisting between the rostrum and the braincase that occurs as a consequence of larger masseter and pterygoid muscles (see section 4.2 in Chapter 4).

Other explanations for the co-occurrence of a lengthened jaw, a diastema, and an anterior vector can be suggested. The size of the teeth in antelopes and others may have changed to some extent, resulting in a somewhat longer or shorter diastema. A long diastema may have an important functional role in some animals. Also, an anterior vector in some rodents could be important to their chewing because of the strong, anteriorly directed jaw movements in some of these animals. Moreover, a longer jaw increases the length of Region II, and that, in turn, increases the bite force at those teeth that are then included in this longer region (see Chapter 2).

However, these considerations do not negate the idea that reorienting the resultant muscle vector to maintain the 3:7 ratio provides a mechanical explanation that deals with a change in the length of the anterior segment of the jaw where a long diastema is produced as a by-product. Nor do they counter the explanation for why a diastema, a lengthened jaw, and an anterior vector occur together in a large number of modern mammals.

Note that the moment arm of the vector cannot be longer than the distance JM in Figure 3.10. This point is of some interest because a very high joint allows a greater increase in the moment arm than a lower joint does. Finally, one would expect the vector to increase its *posterior* inclination if the anterior jaw should *shorten*.

In modern selenodont artiodactyls the 3:7 relationship between the two regions of the jaw has been maintained. Partly because of the above change, virtually all the cheek teeth (premolars as well as molars) lie within Region II, and therefore some premolars have larger bite forces (cf. Fig. 2.5). (At the same time, there are no bite forces where there are no teeth.) Moreover, the cheek tooth rows, like the molar rows in many early members of the group, are almost parallel to each other, suggesting that the forces at these teeth are sub-equal (Fig. 2.5). Further, each cheek tooth row is more-or-less in line with its joint, ensuring a fairly long Region II (a longer region with reasonably high bite forces). Also, the jaw mechanism in these animals presumably functions better when the bite forces are uniform, because small regions of the tooth rows occlude in a wave from back to front during the chewing cycle in these animals.

To summarize, three features of the masticatory apparatus typically occur together in some mammals: (1) a dominant masseter/pterygoid muscle group (resulting in an anterior vector); (2) an excessively long diastema; and (3) a very long jaw. The major result of the previous analysis is that it provides a mechanical explanation for this concurrence. Less clear is which characteristic initiated the other two.

Increased bite forces at some teeth and improved cropping ability are probably the most likely initiators of these features. A longer jaw may improve the ability to pick out particular vegetation in artiodactyls and horses because it makes for a longer and narrower snout, facilitating food choice and collection. This proposition is reasonable, but increased bite forces at extra teeth also seem important. In the case of rodents, a long anterior jaw is a logical necessity when the incisors become very long and grow constantly.

Given jaw lengthening, a long diastema follows necessarily if tooth size does not also increase. Even without an increase in jaw length, if some teeth get smaller (e.g., in bears) a diastema is produced. However, an *extremely* long space produced in this way seems to be uncommon. A long diastema certainly does separate cropping or nipping incisors from grinding/shearing molars in both rodents and selenodont artiodactyls. However, this separation is perhaps unnecessary in modern forms, because it was absent in early artiodactyls. All this implies that a diastema is simply a by-product of jaw lengthening.

Large masseter and pterygoid muscles provide large anterior components of force. These forces no doubt contribute to the wide lateral and medial excursions of the lower jaw in horses and selenodont artiodactyls. However, these larger components do not seem to be absolutely necessary, because mediolateral excursions of the jaw are found in most mammals.

Large anterior components of force presumably contribute to the back-to-front movements of the lower jaw in some rodents. In many other rodents, jaw movements are medially directed, much like those of a typical mammal. Thus, the anterior force components from large masseter and pterygoid muscles appear not to be a requirement for molar grinding by many rodents. However, gnawing with the incisors is important to these animals, so large anterior components are important in that activity.

Thus jaw lengthening, for cropping efficiency and increased bite forces on some teeth in selenodont artiodactyls and horses, and in rodents and rabbits because of the large incisors, is likely to have some importance. Anterior force components may be more important in rodents and rabbits. Again, regardless of what initiated the linkage of a long jaw, a long diastema, and an anteriorly inclined vector, the *major* result of the above analysis is the demonstration that these features are mechanically and logically linked, explaining why they occur together in many cases.

Mammals with anteriorly inclined vectors do not have the two minimum distance pairs seen in animals with posteriorly inclined vectors. This might suggest that these groups have less efficiently constructed skulls. However, the skull in these animals has to resist torsion between the front and back regions of the skull (see section 4.2 in Chapter 4). Therefore, these skulls are efficiently constructed, but the efficiency is achieved in a different way.

3.5.2 *Glyptodon*

Attempting to apply the above analysis to what appears to be a very odd case may be of some interest. Fossils of the South American edentate *Glyptodon* have an extremely high jaw joint that is located not very far behind the most posterior cheek tooth (see Romer's 1966 textbook). Further, the temporal fossa is quite small, a strong dependent flange extends down from the zygomatic arch to well below the level of the tooth row, the angular process on the lower jaw is extensive, and the vertical ramus of the mandible is very long. These features suggest a small attachment surface for the temporalis muscle and large surfaces for the attachment of the masseter/pterygoid complex. Accepting that the masses of the muscles and their orientation have to be inferred in these extinct animals, these features taken together suggest that, in life, this animal almost certainly had an anteriorly directed resultant vector of jaw muscle force. Finally, unlike some edentates in which the braincase and the snout lie more-or-less in line, the snout in *Glyptodon* is bent ventrally, thus forming an angle with the braincase.

As described earlier, using the vertices of triangle JMA (jaw joint, molar, anterior end of the upper jaw) an estimate of the inclination of the muscle vector can be determined by rotating a line that extends from behind the molar until a 3:7 ratio is achieved between the distances from this line to the joint and to the anterior end of the jaw, respectively. The resulting line is generally a reasonable estimate of the inclination of the line-of-action of the muscle vector. Analyzing *Glyptodon* in this manner also indicates that the inclination of the muscle vector was anteriorly inclined, which is also corroborated by the size of the likely muscle attachment sites. Judging from these sites, the masseter and medial pterygoid were very large, whereas the temporalis was fairly small.

The anterior inclination of the muscle vector suggests (following the previous analysis) that at some point in this animal's evolutionary history the anterior jaw lengthened. The masseter and pterygoid muscles, which were anteriorly inclined, presumably enlarged while the temporalis got smaller, thus shifting a previously posteriorly inclined vector (as in some other edentates) to a more anterior inclination. This maneuver presumably maintained the 3:7 ratio.

The jaw is likely to have increased in length even though the jaw in *Glyptodon* appears, if anything, to be very short. An analogy with the length of the jaws in cats and dogs is appropriate. We know intuitively what is meant when dogs and cats are said to have long and short jaws, respectively. Yet, when we examine skulls of a dog and a cat with approximately the *same* jaw length – say, those of a small fox and a bobcat – we immediately realize that more information is needed; we cannot very well say that one jaw is longer than the other when they both have the same length. The additional information that is needed is the width of the jaw apparatus. The bobcat is a much wider animal than the fox. Relative to width, the wide bobcat has a short jaw and the narrow fox has a long jaw. Moreover, the mass of the jaw muscles is greater in wide jaws than in narrow jaws, and thus the bite forces are greater. In an analogous way, the jaw in *Glyptodon* only appears to be rather short because it is so high relative to its length.

Nevertheless, the jaw mechanism of this animal seems to be oddly shaped. The following scenario, beyond jaw lengthening and some muscle size change to reorient the vector, can be suggested to explain this morphology. Recall that, in mammals in general, larger muscles are required for increased bite force because the lever system is already optimal. Judging by the very large size of the muscle attachment sites in *Glyptodon*, the muscles may well have been extremely large. This suggests that higher bite forces might have been required *after* jaw lengthening. The ascending ramus, the mandibular angle, and the zygomatic process are all features that could provide the surface area required for the larger muscles.

The strong ventral inclination of the snout is of interest at this point. Compared with some other edentates, this maneuver would lower the anterior end of the tooth row. At the *same* time the *rear* of the tooth row apparently rotated down relative to the anterior end to produce a tooth row that is oriented as it is in these animals. Together these rotations would produce the very deep skull that is found in *Glyptodon*. The concomitant elongation of the ascending ramus of the lower jaw was then required to match the repositioning upper tooth row. If more attachment surface for the masseter/pterygoid complex were necessary, a further increase in the size of the angular process and further lengthening of the descending jugular process would provide such an attachment surface.

The result is an almost "square" masticatory apparatus that is very high relative to its length. Again, the high skull and jaw only makes the jaw appear to be very short because the dorsoventral dimension is much greater than in other edentates and in most mammals in general. Rather than a diastema, teeth are absent anteriorly.

3.5.3 Entelodonts

Entelodonts were a group of fossil artiodactyls that had a jaw joint that was on approximately the same level as the teeth (see Romer's 1966 textbook). Moreover, judging from the muscle attachment sites (especially the long, dependent zygomatic process), a muscle vector that was close to forming a right angle with the tooth row seems reasonable. Applying the previous analyses, perhaps the most reasonable account of the development of the jaw mechanism in these animals is as follows.

The jaw lengthened in an animal with only a moderately high jaw joint. To maintain the 3:7 ratio between the edentulous and tooth-bearing regions of the jaw, the masses of the jaw muscles changed and the vector of jaw muscle force rotated forward (see above). In this particular case, rotation of the vector to a near vertical position was sufficient to maintain the ratio. Finally, the jaw joint may have moved

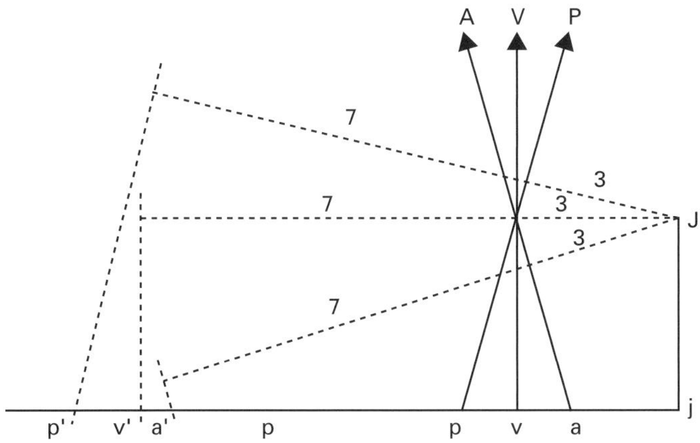

Figure 3.11. A diagrammatic L-shaped jaw with the joint at J, as seen from the side. The corner of the L is at j. Anterior, vertical, and posterior vectors are included on the same jaw to facilitate a comparison. Each vector is located three (perpendicular) units from the joint. The anterior vector intersects the tooth row at the third molar (a) and the anterior end of the jaw is located at a′, which is seven units from the joint when measured perpendicular to the vector. The other vectors have equivalent intersections and distances. After Greaves, W. S. (2000). Location of the vector of jaw muscle force in mammals. *Journal of Morphology*, **243**, 293–299.

closer to the level of the tooth row to secure the wider gape required by the large tusk-like anterior teeth present in these animals.

This case is of particular interest no matter how it originated or the actual inclination of the vector. The vector divides the jaw in the same ratio regardless of its posterior or anterior orientation, because the jaw joint and the tooth row are on essentially the same level (see Fig. 3.1a). Therefore, *any* orientation for the vector gives the same ratio. As expected, the measured ratio is 3:7.

3.6 Estimating vector location from muscle weights and the location of the third molar

As stressed in previous sections, a line drawn *along* a tooth row lies some distance below the jaw joint in most mammals. A perpendicular dropped from the joint (J) down to this line intersects at **j**, at the corner of the L-shaped jaw (Fig. 3.11). Consider the three different muscle vectors (A, V, and P), which are each three (perpendicular) units from the jaw joint. They all are superimposed on the same jaw to facilitate a comparison in Figure 3.11.

The vector that is anteriorly inclined (A) intersects the tooth row rather close to **j** at **a**; the vertical vector (V) intersects three units from **j** at **v**; and the posterior vector (P) intersects relatively far from **j** at **p**. Thus, the distance from the projection of the jaw joint at **j** out to **a**, **v**, or **p** varies with the inclination of the vector. Moreover, each vector is expected to divide its jaw in a 3:7 ratio so that the anterior ends of each jaw are seven perpendicular units from the vector at **a**′, **v**′, and **p**′, respectively. When these seven unit distances are *projected* onto the actual jaw (**aa**′, **vv**′, or **pp**′), these distances also vary with the inclination of the vector.

Accepting that in real animals the intersection of each vector is just behind the third molar, the positions of the intersections are known, because the locations of the third molars are known. Distances **ja** and **ja**′ for the anterior vector are easily measured in a real jaw. Likewise, the equivalent distances for the other vectors, **jv** and **jv**′ and **jp** and **jp**′ can also be measured. Distance **ja** is short compared with **ja**′, so **ja/ja**′ is less than 0.30. The distance **jv** divided by **jv**′ equals 0.30. Finally, **jp** is long compared with **jp**′, so **jp/jp**′ is greater than 0.30. The significance of these quotients is that they indicate the general *inclination* of a vector. The quotient for an anterior vector is less than 0.30, for a vertical vector is equal to 0.30, and for a posterior vector is greater than 0.30.

A second, although rough, indication of the inclination of the vector is available from the weights of the jaw muscles. If the temporalis is large compared with the masseter and pterygoid taken together, the inclination of the vector is very likely to be inclined to the rear. If the masseter and pterygoid are together larger than the temporalis, the vector will probably be anteriorly inclined.

Thus, there are two *independent* indications of the inclination of the muscle vector. First is the geometric idea that the lines-of-action of the resultant vector of jaw muscle force intersect the tooth row at different locations. Second is the muscle weight. These two indications should generally agree and thus provide an approximate test of the idea that the vector intersects the tooth row just behind the third molar.

If the muscle weights suggest a posterior vector, the quotient **jp/jp**′ should be relatively large, and its magnitude will be greater than 0.30. The quotient **ja/ja**′ should be small and be less than 0.30 if the muscle weights indicate an anterior vector. Fewer vectors are expected to be vertical, but they will have a quotient (**jv/jv**′) equal to 0.30.

Weights of the three large jaw muscles are available for about 45 animals from a number of different major mammalian groups. The inclination of the individual muscles is not known. Therefore, subtracting the smaller from the larger muscle group and dividing by the total weight of all three muscles gives a fraction that indicates whether the vector is strongly or weakly inclined either posteriorly or anteriorly. Multiplying by 100 gives a number between 0 and 100. Anterior vectors are assigned a negative sign. Measurements along the tooth row were taken from skulls of the same species to provide the quotients, because this information is not available from the original dissections.

The graph shown in Figure 3.12 can be produced with these weight numbers on the y-axis and the quotients derived from dividing the distance along the actual jaw from **j** to the third molar by the distance from **j** to the anterior end of the jaw on the x-axis.

Points in the upper right region of the graph (greater than 0.3. on the x-axis and positive on the y-axis) represent animals with a posterior inclination according to

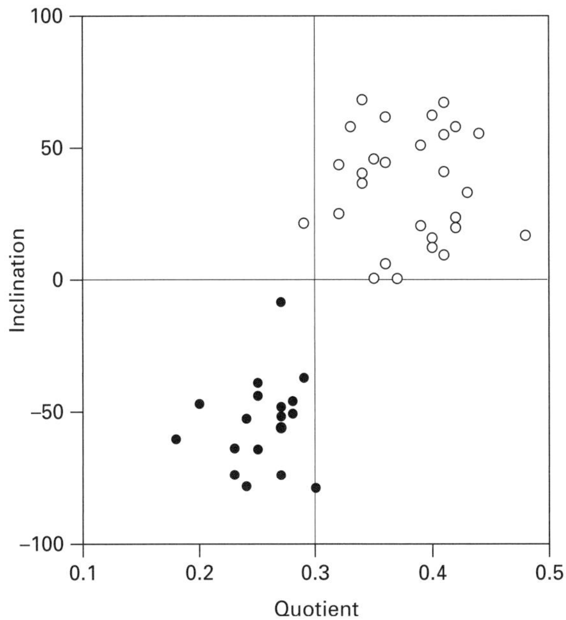

Figure 3.12. A graph in which the inclination of the muscle vector derived from muscle weights is plotted against the quotient derived from the distance from j to the third molar divided by the distance from j to the projection of the anterior end of the jaw (cf. Fig. 3.11). Open circles indicate species in which both measures indicate that the muscle vector is posteriorly inclined. Filled circles indicate species in which both measures indicate that the muscle vector is posteriorly inclined. After Greaves, W. S. (2000). Location of the vector of jaw muscle force in mammals. *Journal of Morphology*, **243**, 293–299.

both the muscle weight data and the values of the quotient. Points in the lower left region (less than 0.30 on the *x*-axis and negative on the *y*-axis) represent animals with an anterior inclination according to both measures. Points for real animals are expected to be located in both of these regions.

Points in the upper left region of the graph represent animals with a *posterior* inclination according to the weight data and an *anterior* inclination according to the values of the quotient. Points in the lower right region represent animals with an *anterior* inclination according to the weight data and a *posterior* inclination according to the values of the quotient. Points for real animals are not expected in either of these areas of the graph.

Virtually all the points representing animals for which weight data are available are found where they are expected to be located. Only one point is found out of place, and this point represents an atypical animal (the giant panda). Few of the points on the graph lie where the quotient is 0.30, and few are expected there because few mammals have a vertical vector. These data are not robust enough to say that the vector does intersect the tooth row just behind the third molar, but nevertheless they do seem to support this idea.

3.7 Summary

The inclination of the resultant vector of jaw muscle force varies directly with the height of the glenoid cavity (on the skull) above the level of the upper tooth row. A very high jaw joint is paired with a vector inclination that approaches the vertical, whereas a very low joint is associated with a vector that is inclined strongly to the rear, at about 50°. The line of action of the muscle force vector is restricted to a region about 5° to either side of a line that makes an angle of 42° with the line from the jaw joint to the rear of the third molar.

Vectors that are inclined anteriorly toward the front of the skull have evolved in relatively few groups of modern mammals (antelopes and deer, rodents, and rabbits), but these groups include a large number of species. In these groups the anterior part of the jaw, in front of the muscle resultant force, became longer either because this part of the jaw itself lengthened or because enlarged teeth made this anterior jaw region longer. Such a change in length could potentially change the 3:7 ratio between the perpendicular distances from the line-of-action of the muscle vector to: (1) the joint; and (2) the anterior end of the jaw. However, simple concomitant changes in the relative masses of the major jaw-closing muscles shifted the line-of-action of the muscle vector anteriorly so that it became directed up and more anteriorly. These changes altered the perpendicular distances from the vector to the joint, and from the vector to the end of the jaw, thus maintaining a 3:7 ratio.

A study of the weights of jaw muscles is used to estimate the location of the resultant vector of jaw muscle force. These weights are combined with the idea that posterior vectors intersect a line along the tooth row, just behind the third molar, farther from the projection of the jaw joint than anterior vectors. Even though the study uses a small amount of less robust muscle weight data, the sample is diverse and suggests that the vector does indeed intersect the jaw just behind the third molar at 30% of the way along the jaw.

4 Skull torsion and the postorbital bar

In cylindrical structures subjected to torsion, maximum compressive and tensile shear strains lie along helices on the surface that form an angle of 45° with the long axis of the cylinder. These two helices intersect at an angle of 90°. The general structure of mammalian skulls approaches one-half of a right circular cylinder, where the flat side represents the palate. As a result of chewing or biting on only one side of the head at a time, the anterior rostral region of many mammalian skulls tends to be twisted relative to the posterior braincase, around an anteroposterior axis. Helices as described above can generally be traced from the braincase to the rostrum on the surface of these skulls.

In many modern selenodont artiodactyls (antelopes, deer, etc.) the postorbital bars contribute to such buttresses: they lie along helices that form an angle of 90° with each other and 45° with the long axis of the skull, modeled as a half cylinder. These helices lie at the surface of the skull as far from the twisting axis as possible for maximum effect, and they connect the braincase and the rostrum by bridging the inherently weak orbital region. A less obvious preorbital tract also meets these requirements.

Torsion in carnivore skulls results when the canine tooth on one side meets resistance while the other encounters soft tissue. Compressive and tensile helices that connect the canine tooth with the opposite jaw joint can be traced on the surface of these skulls. The most efficient connection between the braincase and the anterior region of the skull requires that the ratio between the length and width of the skull be less than or equal to $\pi/2$. Carnivore skulls meet these requirements, and in no case is the ratio greater than $\pi/2$.

4.1 Skull torsion

Many, if not most, mammals chew with the teeth on only one side of the head at a time. The teeth on the non-biting side do not come into contact. Some of the previous analyses in this book have contended with the presence of this unilateral biting, which is so common in mammals. The lower jaw is pulled toward the skull by the great jaw-closing muscles. Only three points, or small regions, of potential contact between these two elements of the head are present. The first of these regions is naturally where the upper and lower teeth, or the food that is positioned between them, meet. The production of bite force at this region is a major function of the jaw mechanism. The remaining areas of contact are located at the two jaw joints. The force generated at these joints is typically transmitted through a fibrocartilaginous disk that separates the two surfaces of the joint and divides the cavity of the jaw joint into two chambers.

The active tooth row is heavily loaded at the bite point, but the teeth on the opposite side naturally resist no force. The forces at the jaw joints on each side of the head are also very different. There are high forces on one joint and lower, to very low, forces on the other. Moreover, on the side where the tooth forces are high, the joint force is low. On the opposite side, the tooth forces are zero and the joint force is high. Thus, a tooth row and the joint, on diagonally opposite sides of the head, resist high forces. The other tooth row resists no forces and its diagonally opposite joint resists low, or even very low, forces. At the same time, there are resultant muscle forces at each side of the head, often with higher forces on the chewing side. As a consequence of this arrangement, the skull is loaded asymmetrically. The front, or rostral part of the skull is twisted in one direction, whereas the rear part, the braincase, is twisted in the opposite direction. Modern antelopes and their kin are among the many mammals that chew on only one side of the head at a time and are an especially good example of this kind of torsion, or twisting.

Suppose biting is taking place at teeth that are located at B on the left side of the jaw in Figure 4.1. Further, suppose that this tooth is located at the end of a line that extends from the right jaw joint (Jr) and passes through the midline at a distance of three units from the level of the jaw joints. In such a situation, the highest reaction forces are found at the bite point (B) and the right jaw joint (Jr), which is located on the opposite (right-hand) side of the head (cf. Figs 2.2 and 4.1). The resultant muscle force (M) on the *chewing* side is located at the middle of one side of the triangle (JrBJl), whereas the resultant muscle force (M) on the *non-chewing* side is some distance away from this triangle. In this idealized case the hypothetical triangle-of-support (JrBJl) collapses to a line (JrB). Thus, the left joint (Jl), on the chewing side, resists essentially no reaction force. This situation was discussed in some detail in Chapter 2.

The skull in this circumstance can be likened to a rectangular plate or a shallow rectangular box (Fig. 4.1). The two jaw joints lie at the corners of one of the short sides. The bite point (B) is located on one long side, six units from the jaw joints. The muscle vectors (M) are each located three units from the jaw joints on each long side. The tooth row (Tr) on the non-biting side extends from M to the anterior end of the rectangle and is free of any reaction force.

The bite force at B and the higher joint force at Jr are each directed *upward* at the ends of the diagonal extending across the rectangle, because the lower jaw is pulled up by the jaw muscles and therefore pushes up on the skull at the right joint and the bite point. The muscle forces (M) are directed *downward* because the muscles pull down on the skull (as well as up on the lower jaw). Such a force regime tends

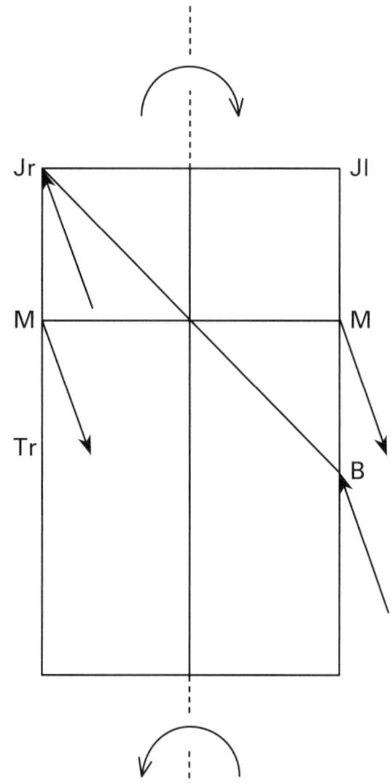

Figure 4.1. The upper jaw represented as a rectangular plate. The muscle forces (M) are directed downward. The forces at the right joint (Jr) and the bite point (B) are directed upward. The left joint (Jl) and the tooth row on the right side (Tr) are unloaded. This loading regime twists the anterior region of the skull in a counterclockwise direction (lower curved arrow) and the braincase in a clockwise direction (upper curved arrow).

to warp the rectangular plate, or box. The corner at the right jaw joint (Jr) moves up slightly because it is heavily loaded and the left joint (Jl) moves down somewhat because it resists a smaller, or even no, force. The bite point (B) tends to move up by a small amount because it is heavily loaded, whereas the tooth row (Tr) on the opposite side tends to move down because it is unloaded.

Under a loading regime like this, a skull also tends to warp in a similar fashion, although to a much lesser extent. In a frontal view of a skull, this means that the braincase tends to be twisted in one direction (say, clockwise) while the anterior moiety of the skull, the rostrum, is twisted in the opposite (counterclockwise) direction – shown by the curved arrows in Fig. 4.1. Thus, the entire skull is subjected to torsion. The anterior and posterior parts of the skull meet at the orbital region. This region tends to be weaker than some other parts of the skull because the spaces that house the eyes are naturally devoid of any internal structural bone. The strength of the twisting forces on the skull varies with the biting location. Further, when the biting side changes, the direction of twisting reverses.

Engineers have long known that if torsional forces are high enough a cylindrical structure of uniform composition will fracture in a characteristic way. The broken surface has a helical contour, the edges of which can be traced along the outside surface of the structure. These helices form an angle of 90° with each other and an angle of 45° with the long axis of the cylinder. Maximum compressive and tensile shear strains lie along these two separate surface helices. A piece of chalk, twisted to the breaking point, nicely demonstrates these edges. Therefore, material lying along these surface helices is as far from the twisting axis as possible and is best positioned to resist these torsional forces. Of course, skulls are not uniform in composition, nor are they perfect cylinders (or half cylinders), but in the discussion below they will be modeled as half cylinders that are reasonably uniform.

4.2 Selenodont artiodactyls

The selenodont artiodactyls (antelopes, deer, etc.) demonstrate a very clear example of a postorbital bar that extends from the frontal bone to the zygomatic arch. In the skulls of these animals, the angle between the estimated long axis of the skull and the postorbital bars approaches 45°. The angle between the bars themselves is 90°. Therefore, these bars have the same relationship to each other, as do the helices discussed above. This suggests that the bars resist torsion when the anterior and posterior segments of the skull are twisted relative to one another during chewing (Fig. 4.1). On the chewing side the bar is expected to resist compression, whereas the bar on the non-chewing side is resisting tension. The bars are in the best possible positions to resist torsional forces because they are appropriately oriented relative to each other as well as to the long axis of the skull, which is modeled as a half cylinder where the flat side represents the palate. Moreover, these bars are as far from the twisting axis as possible (near the surface of the skull, where the maximum shear strains are located) and connect the anterior and posterior parts of the skull.

Most mammals do not have a postorbital bar. It is found only in the selenodont artiodactyls (antelopes and their kin), primates (monkeys and apes and their relatives), and in a few isolated mammals from some of the other large mammalian groups such as the perissodactyls (horses).

In the modern artiodactyls that have a postorbital bar, two of the major jaw-closing muscles, the masseter and medial pterygoid, are very large. Moreover, they are oriented such that they have large components of force perpendicular to the general long axis of the skull. Thus, these muscles strongly twist the anterior and posterior regions of the skull relative to one another during chewing. The temporalis muscles of these animals tend to be smaller, and their lines-of-action usually meet the axis of the skull at a smaller angle. Therefore, these latter muscles have smaller perpendicular components and twist the skull much less strongly. Mammals that have a large temporalis muscle and smaller masseters and pterygoids tend not to have postorbital bars. In these cases a torsion-resisting postorbital bar is presumably not required, because the perpendicular components of force are smaller (but see section 4.3, below).

In this regard, note that early fossil selenodont artiodactyls (and early horses) did *not* have postorbital bars and appear to have had large posteriorly inclined temporalis muscles. Later animals in these same groups *did* have these bars, together with smaller temporalis muscles and larger anteriorly oriented masseter and pterygoid muscles. These observations are exactly what would be expected. The early animals with large posteriorly inclined temporalis muscles had *lower* twisting force components perpendicular to the long axis of the skull and lacked a postorbital bar. In those animals with large anteriorly inclined masseter and pterygoid muscles, with *high* twisting force components, a bar was present.

[Primates usually have a fairly large temporalis muscle. Smaller twisting forces are expected in primates because the temporalis muscle has smaller twisting components of force. The presence of a large temporalis muscle suggests the *absence* of a postorbital bar in primates. The fact that a bar is *present* in these animals suggests that it is not functioning like the bar in selenodont artiodactyls, in which the temporalis muscle is smaller and the masseter and medial pterygoid muscles are larger. In primates, the postorbital bar, plus a wall of thin bone, forms a bony socket for the eye. Some have suggested that in most primates the main function of this bony socket (including the postorbital bar) appears to be to isolate the eye from the jaw muscles. Thus, stereoscopic vision is not disturbed during chewing.]

In addition to the postorbital bars, there are two much less obvious helical tracts that are present in selenodont artiodactyls. Like the postorbital bars they are situated far from the axis of rotation, for maximum effect, and lie at the surface of the skull along lines coincident with the maximum tensile and compressive shear strains that lie at 45° to the long axis of the skull and connect the anterior and posterior portions of the skull. They also lie at 90° to each other.

This second pair of tracts or helices is much less obvious because they are not freestanding struts like the postorbital bars. Rather, they are simply areas of the skull surface, which need not be thickened, where it is possible to trace a *continuous* helical line. This helical tract begins at one jaw joint, passes behind the orbit on the same side, crosses the frontal bone on the midline of the skull, lies in front of the orbit on the opposite side, and finally ends near the most anterior cheek tooth, avoiding the facial vacuities that are sometimes present in the skull of these animals. This helical tract intersects both the braincase and the anterior tooth-bearing segment of the skull. Each postorbital bar and the anterior tract that accompanies it are essentially parallel to each other. The former passes behind, and the latter in front of, the orbit. Recall that the large empty (of bone) orbits are largely responsible for the weak area between the anterior and posterior skull regions.

One could argue that this second buttress-like region could be imagined on any structure, like the skull, that has a shape that approaches a cylinder (or a half cylinder). This is certainly true, but the important point is that such a path can be traced *continuously* on the complicated outer surface of the skull, in the appropriate orientation (lying at 45° to the long axis of the skull and 90° to the path of its opposite on the other side of the skull), far from the axis of rotation where it will have the most effect, and connecting the anterior and posterior parts of the skull.

Interestingly, teeth are generally not found anterior to where this second tract intersects the tooth row. Such teeth would require extra bony buttressing if they were joined to the braincase via a helical tract (cf. section 4.3).

[The orientation of more anterior "bars" and tracts in rodents and rabbits is suggestive, in that arguments similar to the above could be advanced. However, these animals have not yet been studied adequately.]

4.3 Torsion in the skull during canine biting

Often during biting, one of the large canine teeth at the front of the jaw in the Carnivora will encounter resistance (a bone, say) while the other meets only soft tissue. At the same time the jaw joint diagonally opposite the loaded canine may be more heavily loaded than the other joint during the beginning of the bite. In such a case, the skull is initially loaded asymmetrically. The rostrum rotates in one direction as the unloaded canine moves down. At the same time the braincase rotates in the opposite direction as the jaw joint with the lower force moves down and the remaining joint, with somewhat higher forces, moves up. The entire skull is twisted around an approximately anteroposterior axis. Subsequent changes in muscle activity may lessen this twisting, but initially the skull resists large torsional forces.

As noted earlier, the Carnivora have smaller components of force acting perpendicular to the long axis of the skull. Therefore, the torsional forces that twist the two halves of the skull are smaller and *postorbital* bars are absent. Nevertheless, these animals exert very strong forces when biting with their canine teeth. Thus, strong muscle forces, even at a less than optimal orientation for torsion, still produce substantial twisting forces.

Consider a typical dog skull. Two helices that wind around the skull can be imagined. Each helix begins at a canine and ends at the diagonally opposite jaw joint. These two helices intersect at an angle of 90° and each meets the long axis of the skull at 45°. When the skull is twisted, compression will be resisted along the helix beginning at the loaded canine. Tension will be resisted by the perpendicular helix.

If these helices, which begin at one canine tooth and wind around the skull to the diagonally opposite jaw joint, are examined carefully in different carnivores, three possibilities exist. First, and most importantly, the helix can intersect the jaw joint on the opposite side of the skull, as in dogs. Second, the helix can intersect behind this joint. Third, the helix can intersect in front of this joint.

In the first case, as suggested above, the maximum torsional forces lie along the helix that extends from one canine to the diagonally opposite jaw joint. The half-cylindrical structure of the skull may be sufficient, but extra bone tissue, or folds and curved surfaces in the skull, can be deposited or located along these helices if needed to resist twisting the braincase relative to the rostrum.

In the second case, the helix emanating from the canine, although not intersecting the joint, will still intersect well within the posterior part of the skull. Moreover, a helix beginning at the jaw joint will not intersect the canine, but it will intersect well within the rostral region of the skull. Therefore, in this case *two* helices are resisting skull torsion. This skull, because it has two helices, is overly strong compared with the first case when biting produces a load on only one canine.

The third possibility is very different from the first two. In this case the helix from the canine will not intersect the posterior moiety of the skull. The helix from the jaw joint will intersect only the posterior part of the rostrum far behind the canine. A helix beginning at the *canine* is not sufficient to resist skull torsion, because it does not form a connection between the points on the front and rear portions of the skull. The helix beginning at the other *joint* does not connect the forces at the front and rear portions of the skull either. Because these helices are not coincident, there is an interval between them. Unlike the other two cases, this interval would have to be bridged with additional bone tissue before these helices could effectively resist torsion.

As before, imagine that the skull is represented as a half cylinder, where the flat side is the palate. A helix begins at one corner at the end of the half cylinder where the curved and flat surfaces meet (the canine) and winds around the curved surface to the opposite end, where it intersects the diagonally opposite corner (the jaw joint). The helices meet the edge where the curved and flat surfaces meet at an angle of 45°. Further, imagine that the curved surface of the half cylinder is unrolled and laid flat to form a rectangle. The (now straightened) helices are straight lines that bisect the corner angles (forming two 45° angles at each of the four corners). Therefore, the rectangle is a square and the helices are its diagonals. One side of the square is the linear distance from the joints to the canines. The other side is the (unrolled) circumference of the half cylinder.

Calculations that take into account that the helices form an angle of about 45° with the long axis of the skull indicate that, in the case where one helix intersects both the canine and the joint, the length of the jaw divided by its width is equal to $\pi/2$. Length is the anteroposterior distance measured along the midline from the level of the canines to the level of the joints. Width is the distance between the two jaw joints.

This half circumference is equal to $\pi D/2$, where D is equal to the width (w) of the jaw from joint to joint. Thus, this half circumference is equal to $\pi w/2$. As half the circumference also equals the length of the half cylinder from canines to joints (the flattened half cylinder is a square), then $\pi w/2$ = length (l). After rearranging the terms, l/w is equal to $\pi/2$.

In a relatively short skull, *two* helices intersect the posterior and anterior skull regions and the length-to-width ratio is *less* than $\pi/2$. In a skull of medium length, the helix intersects both the jaw joint and the canine on the opposite side. In this case the length divided by the width is equal to $\pi/2$. In the case where neither helix intersects both the anterior and posterior regions of a very long hypothetical skull, the ratio is *greater* than $\pi/2$.

These considerations suggest that the length-to-width ratio of carnivore skulls, and presumably other mammalian skulls that are subjected to this kind of torsion, will be equal to, or less than, $\pi/2$. Skulls longer than this are not expected in nature because they require extra bone tissue or architectural change to resist torsion. Dog-like skulls with a length-to-width ratio of $\pi/2$ will adequately resist torsion by virtue of their half-cylindrical shape (or if necessary, with minor architectural change or a small amount of extra bone tissue located along each helical tract). Very short cat-like skulls, with a ratio less than $\pi/2$, will be able to resist fairly large torsional forces, because these forces will be concentrated along two helices rather than just one. Extra bone tissue along two helices can potentially produce a very strong skull.

One can therefore predict that: many carnivores will have skulls with a length-to-width ratio of $\pi/2$; none will have a ratio greater than this; and animals such as cats, which have larger jaw-closing muscles that produce larger torsional forces, will have very short skulls that are able to resist these greater twisting forces because of the presence of two helices.

A sample of animals, from the major families of the Carnivora, has been measured. The length-to-width ratios of the jaw mechanism in canids (dogs), viverrids (mongooses and related forms), and ursids (bears) do not significantly differ from $\pi/2$. Felids (cats), mustelids (minks, otters, etc.) and hyenas, with larger jaw muscles and greater torsional strains, have the shortest jaws with a ratio less than $\pi/2$. The remaining carnivore families have intermediate ratios. No carnivore jaws have ratios greater than $\pi/2$.

The above suggests that animals such as dogs have jaws that are as long as they can safely get without requiring excessive architectural change or extra bone tissue. Cats have short jaws. They also have larger bite and torsional forces that can be handled by their overbuilt skulls (if necessary, bone tissue can be distributed along *two* helices, producing even greater torsional resistance than is seen in dogs). There are no carnivores with jaws that are "too" long, because presumably they would require too much extra bone tissue. These measurements suggest that minimizing the amount of bone tissue that is required to resist torsion is an important issue in the structure of the carnivore skull, and probably other mammalian skulls as well. Some of the previous analyses in this book also suggested the importance of minimizing the amount of bone tissue used in the construction of the mammalian skull.

Now imagine selection for larger bite forces in a hypothetical carnivore. Increasing the animal's size increases its muscle mass and therefore its bite force. Alternatively, extra muscle mass alone could be added, but that would change the basic shape of the animal (a dog-like skull becoming *wider* to accommodate more muscle tissue becomes more cat-like). In an evolutionary sense, simply increasing size might be more likely than changing shape. This perhaps explains why plotting length vs. width of skulls in different carnivore families produces a series of straight lines, where the points derived from each family cluster around a particular straight line. Rough indicators such as length and width are expected to give similar (but not exactly the same) ratios in animals that are in a sense just different-sized "copies" of each other. Cats are a good example; they are all remarkably similar except for size.

The discussion above may explain why the slopes of the lines for each family are as close to each other as they are: the length-to-width ratio for most carnivores is $\pi/2$ or slightly lower, whereas just a few groups can be said to have a very low ratio. Some mustelids with a relatively long, low skull may simply *appear* to be exceptions. Note that $\pi/2$ refers to *jaw* dimensions (joint to canine vs. joint to joint), not *skull* dimensions.

Recall the previous discussion about the anterior helical tract in selenodont artiodactyls. The distance along the midline from the level of the jaw joints to the anterior premolars is in some ways comparable to the distance from the level of the joints to the canines in carnivores. In both cases the width of the skull (joint to joint) is related to the distances from the joint to the premolar in artiodactyls and to the canine in carnivores. The length-to-width ratio, for these parts of the artiodactyl skull, should approach a value of $\pi/2$. This is true because that is the distance required for the helix, beginning at one joint, to pass around the (roughly) half cylindrical skull to the premolar on the tooth row on the opposite side. This configuration is the most efficient, as alluded to above, and is what is generally found in the artiodactyls.

Recall also the previous discussion concerning the diastema in selenodont artiodactyls. In a lengthened jaw, where the teeth did not increase appreciably in size, a diastema is a necessary outcome. Nevertheless, the previous paragraph suggests that there are at least two reasons for a very long diastema. First, a portion of the diastema is caused by a lengthened jaw. Second, a smaller increase in the length of the diastema may be because of the loss of at least one tooth that was located in front of the anterior helical tract.

4.4 Summary

Unilateral chewing subjects many mammalian skulls to torsional forces that twist the braincase in one direction and the rostral region in the other, around an approximately anteroposterior axis. The maximum shear strains in a structure that is twisted like this lie along helices that wind around the axis of rotation, and the strains are largest far from this axis. These helices meet each other at an angle of 90° and meet the long axis of the skull at an angle of 45°. The generally tubular or half-cylindrical shape of many mammalian skulls is a shape that can resist torsional forces relatively well. However, at inherently weak areas, such as the orbital region, which connects the front and back regions of the skull, buttressing or architectural changes may be required.

The selenodont artiodactyls display two features that can augment the resistance to torsion. The postorbital bars and their extensions are located along helices as described above. In addition, preorbital "helices" can also be traced around the slightly curved surfaces of the skull. These preorbital tracts (which are parallel to the postorbital bars) begin at the jaw joint on the braincase, pass in front of the orbit on the opposite side, and finally intersect the anterior end of the cheek tooth row.

Torsion in carnivore skulls occurs when one canine tooth meets (bony) resistance on one side but not on the other. Maximum compression and tensile forces are found along helical tracts that wind around the skull, modeled as a half cylinder, and meet its long axis at 45°. They meet each other at an angle of 90°. The length-to-width ratio of the skull is important in resisting this torsion.

There are three basic possibilities. In relatively long, narrow, dog-like skulls a single helix connects the front and rear parts of the skull. In relatively short, wide, cat-like skulls two helices connect the two halves of the skull. An overly strong skull is possible in this case. Finally, in an *overly* long, narrow *hypothetical* skull, two helices plus extra bone to bridge the gap between them are required to connect the two portions of the skull.

Most carnivores have long, narrow skulls where one helix is evident. In some groups, such as cats and hyenas, where an overly strong skull might be expected, the skull is found to be relatively short and wide, and *two* helices are evident. Not surprisingly, the very inefficient case in which the skull would be overly long and narrow and where a helix would *not* intersect both the braincase and the rostrum is not found in carnivores. The preorbital helices, observed in selenodont artiodactyl skulls are in some ways equivalent to the helical tracts found in carnivore skulls.

General summary

The jaw viewed as a two-dimensional lever

At a basic level, the jaw mechanism of mammals is remarkably similar in most cases. Simple mechanical models are used to suggest what this natural design of the jaw mechanism is and, therefore, why the jaw mechanism is constructed as it is. [See section 1.1]

Initially, the jaw is modeled as a single straight line with the jaw joint located at one end, the muscle force located somewhere in the middle, and the tooth row extending back from the other end. As additional simplifications, only one side of the jaw is considered and the muscle forces are taken to be a single unit vector that is perpendicular to the line representing the jaw, a construct that greatly simplifies the arithmetic. This model, although extremely simple, still conforms to what is generally observed in animals. [See section 1.1]

Maximum bite forces were assumed to be important as a first approximation. This first assumption failed, as real animals looked nothing like that predicted by the model: muscles located as far forward as possible and teeth positioned as close to the joint as possible. [See section 1.4]

Even so, the model, simple as it is, does appear to be generally correct. After further study, what eventually became clear was that the forces at the jaw joint have to be considered *first*, before examining the bite forces. Study of the relative joint forces when biting at each point along the jaw clearly indicates that the location of the bite point determines whether the joint forces will be compressive or tensile. [See section 1.5]

Compressive forces are acceptable because mammalian joints, such as the jaw joints, are compression-resisting devices. Tensile forces would pull the joint apart and require significant anatomical structures (very strong joint ligaments, large buttresses of bone, or strong appropriately placed muscles) able to resist what can be very high forces. Yet such anatomical structures are usually absent. [See section 1.5]

However, the complete elimination of tensile forces simply requires that biting take place only in those locations that produce compressive forces at the joint. These locations are those that are anterior to the line-of-action of the muscle resultant force. Restricting teeth to these locations completely eliminates tensile forces at the jaw joint. [See section 1.5]

Yet this restriction calls attention to another difficulty. If all the teeth are anterior to the estimated muscle force, the length of the tooth row then varies with the location of this force. When the muscle vector is far forward, the length of the tooth row is necessarily very short. It can be very long when the vector is far to the rear. Moreover, a muscle vector located far forward produces high bite forces, while one located far to the rear produces low forces. [See section 1.6]

Up to this point, bite forces were considered at only one location at a time. These forces are low anteriorly and high posteriorly. However, over time an animal will use all of its teeth, so the *sum* of the forces, at all of the teeth, may be important. At this point one could legitimately ask: Are a few high forces better or worse than many low forces? Summing the bite forces produced by the muscle force when it is positioned at different locations along the jaw demonstrates that these sums vary with the location of the muscle vector. These sums are low when the vector is farther toward the rear, as expected because the leverage is poor. However, they are also low when the vector is close to the front of the jaw. The bite forces here are high, but there are fewer of them. The highest *sum* of bite forces is evident when the muscle vector is located about one-third of jaw length from the joint. This location is approximately that found in all cases when the location of the muscle vector is estimated by careful dissection and weighing of the jaw muscles in a diverse sample of mammals. [See section 1.7]

In all of these cases the relevant distances are measured, as in all lever systems, perpendicular to the line-of-action of the muscle vector, which is the input force in this case. Presumably, this approximately one-third relationship generally went unnoticed because most researchers do not routinely attempt to estimate the line-of-action of the jaw muscle force: first, because it would be only an estimate; second, because of the inherent difficulties associated with this kind of estimation; and third, because of the concern that muscle attachments may not be a good guide to mechanical advantage. Nevertheless, given that the jaw is a type of lever, knowledge of the line-of-action is required in order to calculate relative bite forces.

As interesting as these results are, they refer to a two-dimensional jaw. Naturally three-dimensional models must be studied, but again, keeping them as simple as possible.

The jaw viewed as a three-dimensional lever

Avoiding tensile forces in the two-dimensional case requires only that all the teeth be located in front of the line-of-action of the muscle force. In a three-dimensional case with two active jaw joints and two sets of muscles, tensile forces are avoided by keeping the muscle force inside the triangle formed by the two jaw joints and the bite point. This triangle-of-support can be likened to a three-legged stool that remains stable as long as the weight on the seat is inside the triangle formed by the three legs. Shifting the weight outside this triangle tilts the stool so that one leg moves away from the floor. The jaw behaves in a similar manner (although upside down). When the muscle force lies outside the triangle-of-support, one jaw joint is disarticulated, or pulled apart. As noticed earlier when studying the two-dimensional case, there are no anatomical structures that can resist these tensile forces if they are too high. Thus, eliminating tensile forces simply means keeping the presumed line-of-action of the muscle force inside the triangle-of-support. [See section 2.2]

The jaw is a triangular plate in this three-dimensional model. When biting with anterior teeth, the line-of-action of the muscle force is almost always located inside the triangle-of-support. When biting far to the rear of the tooth row, the line-of-action falls outside this triangle. Thus, adjustments must be made to avoid this situation. Shifting the line-of-action to either side so that it remains inside the triangle is accomplished by reducing the muscle force on the non-biting side of the head. This means that *less* input muscle force is available. Because less muscle force is put into the jaw lever system, lower bite forces are produced at the posterior end of the tooth row. They are lower when they are compared with posterior bite forces in the two-dimensional model, even though the leverage is much better at the back of the jaw. [See section 2.5]

Perhaps the most significant finding from the three-dimensional model, to this point, is that the location for the muscle line-of-action that maximizes the sum of the bite forces is a little less than one-third of jaw length from the jaw joint. This best location for the line-of-action in the three-dimensional case turns out to be 30% of the way along the jaw. That is, the distance from the joint to the vector is three units and the distance from the vector to the anterior end of the jaw is seven units. Thus, the vector divides the jaw in a 3:7 ratio when measured perpendicular to the vector. [See section 2.6]

The sum of the bite forces also varies with the mediolateral location of the tooth row (with the muscle vector always dividing the jaw in a 3:7 ratio). When jaw length and jaw width are taken into account, there is a location for the tooth row at which the sum of the bite forces has a maximum value. In a sample of mammals, the actual location of the tooth row matched the location where this sum is the largest. [See section 2.9]

In a lateral view, how the upper and lower tooth rows meet each other is dependent upon the perpendicular distances from each row to the jaw joint. If the distances are different, or if the joint is above one row and below the other, the tooth rows meet at some angle. If the distances are the same, the joint can be above, below, or on the same level as the tooth rows. The rows will then meet simultaneously all along their length. Put differently, simultaneous occlusion requires that the lower tooth row, together with the jaw joint, can be superimposed on the upper tooth row and the joint. [See section 2.10]

When teeth are worn sufficiently, enough enamel is worn away to expose the dentine. The topography of the surface of the resulting wear facet allows the determination of the relative motion between two occluding teeth in those many animals that habitually move the lower jaw in a single direction during chewing. The interface between the enamel and the dentine is flush on the side of a tooth that encounters the occluding tooth first (the leading edge). A step forms between the dentine and enamel at the interface on the opposite side of the tooth (the trailing edge). [See section 2.11]

Vector inclination and joint location

In all the previous models, the line-of-action of the muscle vector was positioned perpendicular to the line representing the jaw. This relationship simplified some elementary calculations. The line-of-action in real animals is usually inclined toward the rear, and less often inclined toward the front, of the jaw. [See section 3.1]

Considering posterior inclinations first, they vary directly with the height of the jaw joint above the level of the tooth row. A vector inclination that approaches the vertical is associated with a very high jaw joint. A vector that inclines about 45° to the rear is paired with a jaw joint that is only a short distance above the tooth row.

Consider a line drawn from the jaw joint to the third molar in lateral view. Now imagine a second line that is anterior to the first, that begins at the molar, and where the angle between the two is 42°. In general, the line-of-action of the muscle force is located about 5° to either side of this second line. If the jaw joint is relatively high above the level of the tooth row, the line at 42° will point up and slightly toward the front of the jaw. If the joint is relatively low, this line will point up and toward the rear. [See section 3.4.4]

Anteriorly inclined vectors are less common in mammals and are usually found in selenodont artiodactyls (antelopes, etc.), rodents, and lagomorphs (rabbits). The region of the jaw anterior to the muscle vector has apparently lengthened in these cases. In the case of the selenodont artiodactyls a long diastema necessarily forms; the size of the teeth did not change appreciably, so the teeth no longer fill the jaw. In the case of rodents and lagomorphs, the anterior parts of the jaws presumably are longer simply because the incisor teeth became very long. In none of these cases did this lengthening lead to a change in the 3:7 ratio of the jaw regions. Concomitant changes in the *masses* of the jaw muscles shifted the line-of-action of the muscle vector anteriorly. This anterior rotation of the vector increased its moment arm (as well as the moment arm of the teeth) without any additional changes in the bony parts of the skull or jaw. [See section 3.5.1]

However, a more anterior facing line-of-action of the muscle force meets the long axis of the skull at a larger angle. Therefore, larger components that are perpendicular to the long axis tend to twist the front and back halves of the skull relative to one another. A postorbital bar, which resists this torsion, is usually found in the selenodont artiodactyls with lengthened jaws, but *not* in their ancestors that had muscles with a posterior line-of-action (see below). Anterior bar-like structures are also found in rodents and rabbits but have yet to be studied in detail. [See section 3.5.1]

Anterior muscle vectors intersect a line along the tooth row closer to the projection of the jaw joint than do posterior vectors. Combining these different intersections with the weights of a sample of jaw muscles supports the idea that the third molar is located just in front of the resultant muscle force. Thus, these results support the idea that the vector of muscle force (and the rear of the third molar) divides the jaw in a 3:7 ratio. Dissections and weighing of the jaw muscles as well as the previous models suggest the same thing. [See section 3.6]

Skull torsion and the postorbital bar

The braincase and the rostrum of the mammalian skull are twisted relative to one another in many mammals. Biting on only one side at a time, while using muscles on both sides of the head, produces torsional forces that cause this twisting. Engineers have long known that in a cylindrical body under torsion, the maximum tensile and compressive shear strains lie along surface helices. The angle between the two helices is 90°. The angle between the long axis of the cylinder and both helices is 45°. [See section 4.1]

When a selenodont artiodactyl skull is modeled as a half cylinder, the postorbital bars found in these animals are coincident with the helices described above, suggesting that they may function as struts that resist torsion in the skull. These bars are ideally situated, in that they bridge the distance from the braincase to the rostrum, lie at 90° to each other, lie at 45° to the long axis of the skull, and are at the surface of the skull as far from the axis of rotation as possible, where their effect will be greatest. Moreover, they span the weak orbital region between the braincase and the rostrum. [See section 4.2]

Preorbital "bars" are also present in selenodont artiodactyls. These "bars" or tracts are simply helices that can be traced continuously on the outer surfaces of the skull. They begin at the jaw joint on one side of the braincase, pass forward behind one orbit, then in front of the orbit on the opposite side, avoiding any facial vacuities that may be present, and finally intersect the anterior end of the cheek tooth row. Anterior to this intersection cheek teeth are predicted to be absent because they would require extra bony buttressing, and generally they are not found in these animals. [See section 4.2]

Torsion-resisting helical struts, very similar to the preorbital "bars" in artiodactyls, can also be imagined in carnivore skulls. These tracts pass from the canine tooth to the braincase on the surface of the skull. Torsion in this case results when only one canine meets resistance while the other encounters only soft tissue. In the case of carnivores, there are three possibilities. First, a single helix efficiently connects the canine to the jaw joint on the opposite side of the braincase. Second, two helices connect the rostrum and braincase, producing a much stronger resistance to torsion. Third, a single helix is not long enough to touch both the joint on the braincase and the canine. Therefore, two helices plus bony buttressing between the two are needed to connect the two parts of this hypothetical skull. [See section 4.3]

A length-to-width ratio of the skull of $\pi/2$ is required for the first case and represents a fairly long and narrow dog-like skull. A short and wide cat-like skull, in which the ratio is less than $\pi/2$, represents the second case. Here the two helices can provide extra strength that is needed in animals such as cats and hyenas with relatively large jaw muscles. The third case represents an extremely long and narrow skull, with a length-to-width ratio greater than $\pi/2$. This case is not found in nature because of the need for extra, inefficient buttressing. [See section 4.3]

Finally

The above studies, taken together, explain some of the structural relationships that have evolved in mammalian jaw mechanisms. The major result of these studies is that the majority of mammalian jaws, at a basic level, are remarkably similar if not essentially the same. This similarity is masked by the impressive diversity of many, more superficial, structures. Mechanical efficiency, minimizing the amount of bone tissue used, and the avoidance of mechanical difficulties all lead to this major result.

As an aside

The mandibles of some insects may be mentioned. These structures approximate tetrahedrons that appear as right triangles when viewed from below. The hypotenuse of the right triangle is at the lateral side. The long side meets its opposite on the midline. The short side is posterior. Dimples, which serve as teeth, are located all along the long side and occlude simultaneously with their opposites. The fulcrum of the mandible is located laterally at the junction of the short side and the hypotenuse. The large muscle that rotates the mandible toward the midline in a horizontal plane attaches at the right angle of the triangle via a strong tendon that is perpendicular to the short side. Thus, the length of the short side is the moment arm of the muscle, and bite forces are perpendicular to the long side. The short and long sides of the triangle are expected to be in the ratio of 3:7 if the models discussed above are accepted.

Examination of a plains lubber grasshopper (*Brachystola magna*) and a Jerusalem cricket (*Stenopelmatus fuscus*) indicates that the ratio of the moment arm of the muscle to the length of the tooth row is 3:7. Clearly, the jaw mechanisms of mammals and insects, with jaws of this type, are very different. Nevertheless, this ratio is expected if the muscle force is used most efficiently.

References

Alexander, R. M. (1981). Factors of safety in the structure of animals. *Science Progress (Oxford)*, 67, 109–130.

Ardran, G. M., Kemp, F. H., and Ride, W. D. L. (1958). A radiographic analysis of mastication and swallowing in the domestic rabbit *Oryctolagus cuniculus. Proceedings of the Zoological Society of London*, 130, 257–274.

Avis, V. (1959). The relation of the temporal muscle to the form of the coronoid process. *American Journal of Physical Anthropology*, 17, 99–104.

Avis, V. (1961). The significance of the angle of the mandible: an experimental and comparative study. *American Journal of Physical Anthropology*, 19, 55–61.

Barghusen, H. R. (1973). The adductor jaw musculature of *Dimetrodon* (Reptilia, Pelycosauria). *Journal of Paleontology*, 47, 823–834.

Baron, P. and Dubussy, T. (1980). Biomechanical analysis of the main masticatory muscles in the rabbit. *Journal de Biologie Buccale*, 8, 265–281.

Beatty, J. (1980). Optimal-design models and the strategy of model building in evolutionary biology. *Philosophy of Science*, 47, 532–561.

Becht, G. (1953). Comparative biologic-anatomical researches on mastication in some mammals. I and II. *Proceedings Koninklijke Nederlandse Akademie van Wetenschappen, Amsterdam, Series C,* 56, 508–527.

Beecher, R. M. (1977). Function and fusion at the mandibular symphysis. *American Journal of Physical Anthropology,* 47, 325–336.

Beecher, R. M. (1979). Functional significance of the mandibular symphysis. *Journal of Morphology*, 159, 117–130.

Biknevicius, A. R. (1996a). Functional discrimination in the masticatory apparatus of juvenile and adult cougars (*Puma concolor*) and spotted hyenas (*Crocuta crocuta*). *Canadian Journal of Zoology*, 74, 1934–1942.

Biknevicius, A. R. (1996b). Incisor size and shape: implications for feeding behaviors in saber-toothed “cats”. *Journal of Vertebrate Paleontology*, 16, 510–521.

Biknevicius, A. R. and Ruff, C. B. (1992). The structure of the mandibular corpus and its relationship to feeding behaviours in extant carnivores. *Journal of Zoology, London*, 228, 479–507.

Bowman, K. (1951). The functioning of the temporomandibular joint after extirpation of the disc. *Acta Chirurgerie Scandinavica*, 100, 130–135.

Bramble, D. M. (1978). Origin of the mammalian feeding complex: models and mechanisms. *Paleobiology*, 4, 271–301.

Brehnan, K., Boyd, R. L., Laskin, J., Gibbs, C. H., and Mahan, P. (1981). Direct measurement of loads at the temporomandibular joint in *Macaca arctoides. Journal of Dental Research*, 60, 1820–1824.

Brodie, A. G. (1969). The three arcs of mandibular movement as they affect the wear of teeth. *Angle Orthodontist*, 39, 217–226.

Bryant, H. N. and Russell, A. P. (1992). The position of the carnassials in sabertoothed carnivorans: a test of Greaves's model. *American Society of Zoologists Meeting* (*abstracts*), 140A.

Bryant, H. N. and Russell, A. P. (1995). Carnassial functioning in nimravid and felid sabertooths: theoretical basis and robustness of inferences. In: *Functional Morphology in Vertebrate Paleontology*. ed. J. J. Thomason. Cambridge: Cambridge University Press, pp. 116–135.

Buckland-Wright, J. C. (1978). Bone structure and the patterns of force transmission in the cat skull (*Felis catus*). *Journal of Morphology*, 155, 35–62.

Butler, P. M. (1939). The teeth of the Jurassic mammals. *Proceedings of the Zoological Society of London. (B)*, 109, 329–356.

Butler, P. M. (1952a). The milkmolars of Perissodactyla, with remarks on molar occlusion. *Proceedings of the Zoological Society of London*, 121, 777–817.

Butler, P. M. (1952b). Molarization of the premolars in the Perissodactyla. *Proceedings of the Zoological Society of London*, 121, 819–843.

Cachel, S. M. (1979). A functional analysis of the primate masticatory system and the origin of the anthropoid postorbital system. *American Journal of Physical Anthropology*, 50, 1–18.

Carroll, R. L. (1988). *Vertebrate Paleontology and Evolution*. New York: W.H. Freeman.

Cartmill, M. (1980). Morphology, function, and evolution of the anthropoid postorbital septum. In: *Evolutionary Biology of New World Monkeys and Continental Drift*. ed. R. L. Ciochon and A. B. Chiarelli. New York: Plenum. pp. 243–274.

Colbert, E. H. (1945). *The Dinosaur Book*. New York: McGraw-Hill.

Costa, R. L., Jr. (1977). *Dental Pathology and Related Factors in Archeological Eskimo Samples from Point Hope and Kodiak Island, Alaska*. PhD dissertation. Philadelphia: University of Pennsylvania.

Costa, R. L., Jr. and Greaves, W. S. (1981). Experimentally produced tooth wear facets and the direction of jaw motion. *Journal of Paleontology*, 55, 635–638.

Covey, D. S. G. and Greaves, W. S. (1995). Jaw dimensions and torsion resistance during canine biting in the Carnivora. *Canadian Journal of Zoology*, 72, 1055–1060.

Crompton, A. W. (1962). On the lower jaw of *Diarthrognathus* and the origin of the mammalian lower jaw. *Proceedings of the Zoological Society of London*, 140, 697–753.

Crompton, A. W. and Hiiemäe, K. M. (1969). How mammalian molar teeth work. *Discovery*, 5, 23–34.

Crompton, A. W. and Hiiemäe, K. M. (1970). Molar occlusion and mandibular movements during occlusion in the American opossum, *Didelphis marsupialis* L. *Zoological Journal of the Linnean Society*, 49, 21–47.

Crompton, A. W. and Sita-Lumsden, A. (1970). Functional significance of the therian molar pattern. *Nature*, 227, 197–199.

Currey, J. (1984). *The Mechanical Adaptations of Bones*. Princeton: Princeton University Press.

Davis, D. D. (1955). Masticatory apparatus in the spectacled bear *Tremarctos ornatus*. *Fieldiana: Zoology*, 37, 25–46.

Davis, D. D. (1964). *The Giant Panda: a Morphological Study of Evolutionary Mechanisms*. *Fieldiana: Zoology Memoirs*, 3, 1–339.

DeMar, R. and Barghusen, H. R. (1973). Mechanics and the evolution of the synapsid jaw. *Evolution,* 26, 622–637.

Dessem, D. A. (1985a). *Interactions between muscle activity and jaw joint forces*. Ph.D thesis. Chicago: The University of Illinois.

Dessem, D. A. (1985b). The transmission of muscle force across the unfused symphysis in mammalian carnivores. *Fortschritte Zoologie*, 30, 289–291.

Dessem, D. A. (1989). Interactions between jaw-muscle recruitment and jaw-joint forces in *Canis familiaris*. *Journal of Anatomy*, 164, 101–121.

de Vree, F. and Gans, C. (1976). Mastication in pygmy goats (*Capra hircus*). *Annales Societé Royale Zoologique Belgique*, 105, 255–306.

de Wolf-Exalto, A. E. (1951). On differences in the lower jaw of animalivorous and herbivorous mammals. I and II. *Proceedings Koninklijke Nederlandse Akademie van Wetenschapen, Amsterdam, Series C*, 54, 237–246, 405–410.

Druzinsky, R. E. and Greaves, W. S. (1979). A model to explain the posterior limit of the bite point in reptiles. *Journal of Morphology*, 160, 165–168.

DuBrul, E. L. (1980). *Sicher's Oral Anatomy* (7th edition). St. Louis: C.V. Mosby.

DuBrul, E. L. and Sicher, H. (1954). *The Adaptive Chin*. Springfield: Charles C. Thomas.

Emerson, S. B. and Radinsky, L. (1980). Functional analysis of sabertooth cranial morphology. *Paleobiology*, 6, 295–312.

Erhardson, S., Sheikholeslam, A., Forsberg, C.-M., and Lockowandt, P. (1993). Vertical forces developed by the jaw elevator muscles during unilateral maximal clenching and their distribution on teeth and condyles. *Swedish Dental Journal*, 17, 23–34.

Evans, F. G. (1957). *Stress and Strain in Bones*. Springfield: Charles C. Thomas.

Every, R. G. (1970). Sharpness of teeth in man and other primates. *Postilla,* 143, 1–30.

Fariña, R. A. (1985). Some functional aspects of mastication in Glyptodontidae (Mammalia). *Fortschritte der Zoologie*, 30, 277–280.

Freeman, P.W. (1984). Functional cranial analysis of large animalivorous bats (Microchiroptera). *Biological Journal of the Linnean Society*, 21, 387–408.

Freeman, P.W. (1995). Nectarivorous feeding mechanisms in bats. *Biological Journal of the Linnean Society*, 56, 439–463.

Freeman, P.W. and Lemon, C.A. (1991). Morphometrics of the Family Emballonuridae. *Bulletin of the American Museum of Natural History*, No. 206, 54–61.

Gans, C. (1988). Muscle insertions do not incur mechanical advantage. *Acta Zoologica Cracoviensia,* 31, 615–624.

Gans, C. and de Vree, F. (1987). Functional bases of fiber length and angulation in muscle. *Journal of Morphology,* 192, 63–85.

Gaunt, W. A. (1959). The development of the deciduous cheek teeth of the cat. *Acta Anatomica*, 38, 187–212.

Gentry, A. W. and Hooker, J. J. (1988). The phylogeny of the Artiodactyla. In: *The Phylogeny and Classification of the Tetrapods. Volume 2: Mammals*. Systematics Association Special Volume No. 35B. ed. M. J. Benton. Oxford: Clarendon Press, pp. 234–272.

Gingerich, P. D. (1971). Functional significance of mandibular translation in vertebrate jaw mechanics. *Postilla,* 152, 1–10.

Gingerich, P. D. (1972). Molar occlusion and jaw mechanics of the Eocene primate *Adapis*. *American Journal of Physical Anthropology*, 36, 359–368.

Gordon, J. E. (1968). *The New Science of Strong Materials*. Princeton: Princeton University Press.

Gordon, J. E. (1978). *Structures*. New York: Plenum Press.

Gorniak, G. C. and Gans, C. (1980). Quantitative assay of electromyograms during mastication in domestic cats (*Felis catus*). *Journal of Morphology*, 163, 253–281.

Granger, W. (1908). A revision of the American Eocene horses. *Bulletin of the American Museum of Natural History*, 24, 221–264.

Greaves, W. S. (1971). *Functional anatomy of the merycoidodont head* (*Mammalia, Artiodactyla*). PhD dissertation. Chicago: University of Chicago.

Greaves, W. S. (1973a). Evolution of the merycoidodont masticatory apparatus (Mammalia, Artiodactyla). *Evolution,* 26, 659–667.

Greaves, W. S. (1973b). The inference of jaw motion from tooth wear facets. *Journal of Paleontology*, 47, 1000–1001.

Greaves, W. S. (1974). Functional implications of mammalian jaw joint position. *Forma et functio,* 7, 363–376.

Greaves, W. S. (1978a). The jaw lever system in ungulates: a new model. *Journal of Zoology, London,* 184, 271–285.

Greaves, W. S. (1978b). The posterior zygomatic root in oreodonts. *Journal of Paleontology*, 52, 740–743.

Greaves, W. S. (1980). The mammalian jaw mechanism – the high glenoid cavity. *American Naturalist*, 116, 432–440.

Greaves, W. S. (1982). A mechanical limitation on the position of the jaw muscles of mammals: the one-third rule. *Journal of Mammalogy*, 63, 261–266.

Greaves, W. S. (1983). A functional analysis of carnassial biting. *Biological Journal of the Linnean Soc*iety, 20, 353–363.

Greaves, W. S. (1985a). The mammalian postorbital bar as a torsion-resisting helical strut. *Journal of Zoology, London,* 207, 125–136.

Greaves, W. S. (1985b). The generalized carnivore jaw. *Zoological Journal of the Linnean Society*, 85, 267–274.

Greaves, W. S. (1988a). The maximum average bite force for a given jaw length. *Journal of Zoology, London*, 214, 295–306.

Greaves, W. S. (1988b). A functional consequence of an ossified mandibular symphysis. *American Journal of Physical Anthropology*, 77, 53–56.

Greaves, W. S. (1991a). A relationship between premolar loss and jaw elongation in selenodont artiodactyls. *Zoological Journal of the Linnean Society*, 101, 121–129.

Greaves, W. S. (1991b). The orientation of the force of the jaw muscles and the length of the mandible in mammals. *Zoological Journal of the Linnean Society*, 102, 367–374.

Greaves, W. S. (1993). Reply to Drs. Ravosa and Hylander. *American Journal of Physical Anthropology*, 90, 513–514.

Greaves, W. S. (1995). Functional predictions from theoretical models of the skull and jaws in reptiles and mammals. In: *Functional Morphology in Vertebrate Paleontology*. ed. J. J. Thomason. Cambridge: Cambridge University Press, pp. 99–115.

Greaves, W. S. (1998). The relative positions of the jaw joint and the tooth row in mammals. *Canadian Journal of Zoology*, 76, 1203–1208.

Greaves, W. S. (2000). Location of the vector of jaw muscle force in mammals. *Journal of Morphology,* 243, 293–299.

Greaves, W. S. (2002). Modeling the distance between the molar tooth rows in mammals. *Canadian Journal of Zoology*, 80, 388–393.

Greaves, W. S. (2004). Estimating the line of action of posteriorly inclined resultant jaw muscle forces in mammals using a model that minimizes functionally important distances in the skull. In: *Shaping Primate Evolution*. ed. F. Anapol, R. Z. German, and N. G. Jablonski. Cambridge: Cambridge University Press, pp. 334–350.

Greaves, W. S. (2008). Mammals with a long diastema typically also have dominant masseter and pterygoid muscles. *Zoological Journal of the Linnean Society*, 153, 625–629.

Greaves, W. S. and Covey, D. S. G. (1992). Torsion of the skull due to canine biting and the dimensions of the jaw in carnivores. *Journal of Vertebrate Paleontology*, 12, 31A (Abstract).

Greaves, W. S. and Mucci, R. J. (1997). Mechanical loads in the postorbital bar during mastication in artiodactyls and primates. *American Journal of Physical Anthropology Supplement*, 24, 119 (Abstract).

Gregory, W. K. (1951). *Evolution Emerging*. 2 vols. New York: Macmillan.

Gromova, V. I. (1968). Introduction. In: *Fundamentals of Paleontology*. ed. Y. A. Orlov. Jerusalem: Israel Program of Scientific Translation, 3, 1–585.

Gysi, A. (1921). Studies of the leverage problem of the mandible. *Dental Digest*, 27, 74–84, 144–150, 203–208.

Herring, S. W. (1975). Adaptations for gape in the hippopotamus and its relatives. *Forma et functio,* 8, 85–100.

Herring, S. W. and Herring, S. E. (1974). The superficial masseter and gape in mammals. *American Naturalist*, 108, 561–576.

Herring, S. W. and Scapino, R. P. (1973). Physiology of feeding in miniature pigs. *Journal of Morphology,* 141, 427–460.

Hiiemäe, K. M. (1971). The structure and function of the jaw muscles in the rat (*Rattus norvegicus* L.). III. The mechanics of the muscles. *Zoological Journal of the Linnean Society*, 50, 111–132.

Hiiemäe, K. M. (1978). Mammalian mastication: a review of the activity of the jaw muscles and the movements they produce in chewing. In: *Development, Function and Evolution of Teeth*. ed. P. M. Butler and K. A. Joysey. New York: Academic Press. pp. 359–398.

Hiiemäe, K. M. and Ardran, G. M. (1968). A cineflurographic study of mandibular movement during feeding in the rat (*Rattus norvegicus*). *Journal of Zoology, London*, 154,139–154.

Hiiemäe, K. M. and Kay, R. F. (1972). Trends in the evolution of primate mastication. *Nature,* 240, 486–487.

Hildebrand, M. (1974). *Analysis of Vertebrate Structure*. New York: John Wiley.

Hopson, J. A. (1971). Postcanine replacement in the gomphodont cynodont *Diademodon*. *Zoological Journal of the Linnean Society* (*Supplement No. 1*), 50, 1–21.

Hopson, J. A. and Barghusen, H. R. (1986). An analysis of therapsid relationships. In: *The Ecology and Biology of Mammal-like Reptiles*. ed. N. Hotton, *et al.* Washington: Smithsonian Institution Press, pp. 83–106.

Hopson, J. A. and Radinsky, L. B. (1980). Vertebrate paleontology: new approaches and new insights. *Paleobiology*, 6, 250–270.

Hylander, W. L. (1975). The human mandible: lever or link? *American Journal of Physical Anthropology,* 43, 227–242.

Hylander, W. L. (1978). Incisal bite force direction in humans and the functional significance of mammalian mandibular translation. *American Journal of Physical Anthropology*, 48, 1–8.

Hylander, W. L. (1979a). An experimental analysis of temporomandibular joint reaction force in macaques. *American Journal of Physical Anthropology*, 51, 433–456.

Hylander, W. L. (1979b). Mandibular function in *Galago crassicaudatus* and *Macaca fascicularis*: an *in vivo* approach to stress and strain of the mandible. *Journal of Morphology*, 159, 253–296.

Hylander, W. L. (1979c). The functional significance of primate mandibular form. *Journal of Morphology*, 160, 223–240.

Hylander, W. L. (1984). Stress and strain in the mandibular symphysis of primates: A test of competing hypotheses. *American Journal of Physical Anthropology*, 64, 1–46.

Hylander, W. L. (1985a). Mandibular function and temporomandibular joint loading. In: *Developmental Aspects of Temporomandibular Joint Disorders*. Monograph 16, Center for human growth and development. ed. D. S. Carlson, J. A. McNamara, and K. A. Ribbens. Ann Arbor: University of Michigan, pp. 19–35.

Hylander, W. L. (1985b). Mandibular function and biomechanical stress and scaling. *American Zoologist*, 25, 315–330.

Hylander, W. L. and Bays, R. (1978). An *in vivo* strain-gauge analysis of the squamosal-dentary joint reaction force during mastication and incisal biting in *Macaca mulatta* and *Macaca fasicularis*. *Archives of Oral Biology*, 24, 689–697.

Hylander, W. L., Johnson, K. R. and Crompton, A. W. (1992). Muscle force recruitment and biomechanical modeling: an analysis of masseter muscle function in *Macaca fascicularis*. *American Journal of Physical Anthropology*, 88, 365–387.

Hylander, W. L., Picq, P. G. and Johnson, K. R. (1991a). Function of the supraorbital region of primates. *Archives of Oral Biology*, 36, 273–281.

Hylander, W. L., Picq, P. G. and Johnson, K. R. (1991b). Masticatory-stress hypotheses and the supraorbital region of primates. *American Journal of Physical Anthropology*, 86, 1–36.

Janis, C. M. (1995). Correlations between craniodental morphology and feeding behavior in ungulates: reciprocal illumination between living and fossil taxa. In: *Functional Morphology in Vertebrate Paleontology*. ed. J. J. Thomason. Cambridge: Cambridge University Press, pp. 76–98.

Jerison, H. J. (1973). *Evolution of the Brain and Intelligence*. New York: Academic Press.

Joeckel, R. M. (1990). A functional interpretation of the masticatory system and paleoecology of Entelodonts. *Paleobiology*, 16, 459–482.

Kallen, F. C. and Gans, C. (1972). Mastication in the little brown bat, *Myotis lucifugus. Journal of Morphology*, 136, 385–420.

Kemp, T. S. (1979). The primitive cynodont *Procynosuchus*: functional anatomy of the skull and relationships. *Philosophical Transactions of the Royal Society* (*B*), 285, 73–122.

Kermack, K. A., Musset, F. and Rigney, H. W. (1981). The skull of *Morganucodon*. *Zoological Journal of the Linnean Society*, 71, 1–58.

Kieser, J. A., Gebbie, T. and Ksiezycka, K. (1996). A mathematical model for hypothetical force distribution between opposing jaws. *Journal of the Dental Association of South Africa*, 51, 701–705.

Kirk, E. C. (1930). A study of the dynamics involved in the evolution of the human dentures in the relation of centric occlusion. *Dental Cosmos*, 72, 631–643.

Koolstra, J. H. and Van Eijden, T. M. G. J. (1999). Three-dimensional dynamical capabilities of the human masticatory muscles. *Journal of Biomechanics*, 32, 145–152.

Lafferty, J. F., Winter, W. G. and Gambaro, S. A. (1977). Fatigue characteristics of posterior elements of vertebrae. *Journal of Bone and Joint Surgery*, 59-A, 154–158.

Laithwaite, E. R. (1977). Biological analogues in engineering practice. *Interdisiplinary Science Review*, 2, 100–108.
Landry, S. O., Jr. (1970). The Rodentia as omnivores. *Quarterly Review of Biology*, 45, 351–372.
Lauder, G. (1991). Biomechanics and evolution: integrating physical and historical biology in the study of complex systems. In: *Biomechanics in Evolution*. ed. J. M. V. Rayner, and R. J. Wootton. Cambridge: Cambridge University Press, pp. 1–19.
Leibman, F. M. and Kussick, L. (1965). An electromyographic analysis of masticatory muscle imbalance with relation to skeletal growth in dogs. *Journal of Dental Research*, 44, 768–774.
Marinelli, W. (1933). Die functionelle Bedeutung der Postorbitalspange bei Giraffe und Pferd. *Biologia Generalis*, 9, 201–222. [Cited in Willemse, 1950]
Matthew, W. D. (1901). Fossil mammals of the Tertiary of northeastern Colorado. *Memoirs of the American Museum of Natural History*, 1, 353–447.
Maynard Smith, J. and Savage, R. J. G. (1959). The mechanics of mammalian jaws. *School Science Review*, 40, 289–301.
Mellett, J. S. (1977). Paleobiology of North American *Hyaenodon* (*Mammalia, Creodonta*). *Contributions to Vertebrate Evolution*, 1, 1–134.
Mills, J. R. E. (1966). The functional occlusion of the teeth of Insectivora. *Journal of the Linnean Society, London*, 46,1–25.
Møller, E. (1966). The chewing apparatus. *Acta Physiologica Scandinavica*, 69, supplement 1–229.
Moss, M. L. (1968). Functional cranial analysis of mammalian mandibular ramal morphology. *Acta Anatomica*, 71, 423–447.
Nagel, R. J. and Sears, V. H. (1958). *Dental Prosthetics*. St. Louis: C.V. Mosby.
Naples, V. L. (1982). Cranial osteology and function in the tree sloths, *Bradypus* and *Choloepus*. *American Museum Novitates*, 2739, 1–41.
Naples, V. L. (1985). Form and function of the masticatory musculature in the tree sloths, *Bradypus* and *Choloepus*. *Journal of Morphology*, 183, 25–50.
Naples, V. L. (1987). Reconstruction of cranial morphology and analysis of function in the Pleistocene ground sloth *Nothrotheriops shastense* (Mammalia, Megatheriidae). *Contributions in Science, Natural History Museum of Los Angeles County*, N 389, 1–21.
Naples, V. L. (1995). The artificial generation of wear patterns on tooth models as a means to infer mandibular movement during feeding in mammals. In: *Functional Morphology in Vertebrate Paleontology*. ed. J.J. Thomason. Cambridge: Cambridge University Press, pp. 136–150.
Newman, H. N. and Levers, B. G. H. (1979). Tooth eruption and function in an early Anglo-Saxon population. *Journal of the Royal Society of Medicine*, 72, 341–350.
Noble, V. E., Kowalski, E. M. and Ravosa, M. J. (2000). Orbit orientation and the function of the mammalian postorbital bar. *Journal of Zoology, London*, 250, 405–418.
Norman, D. B. and Weishampel, D. B. (1985). Ornithopod feeding mechanisms: their bearing on the evolution of herbivory. *American Naturalist*, 126, 151–164.
Osborn, H. F. (1918). Equidae of the Oligocene, Miocene and Pliocene of North America, iconographic type revision. *Memoirs of the American Museum of Natural History*, 2, 1–217.
Osborn, J. W. (1993). Orientation of the masseter muscle and the curve of Spee in relation to crushing forces on the molar teeth of primates. *American Journal of Physical Anthropology*, 92, 99–106.

Ostrom, J. H. (1961). Cranial morphology of the hadrosaurian dinosaurs of North America. *Bulletin of the American Museum of Natural History*, 122, 33–186.

Oxnard, C. E. (1971). Tensile forces in skeletal structures. *Journal of Morphology*, 134, 425–436.

Oyen, O. J. (1987). Bone strain in the orbital region of growing vervet monkeys. *American Journal of Physical Anthropology*, 72, 39–40.

Oyen, O. J. and Tsay, T. P. (1991). A biomechanical analysis of craniofacial form and bite force. *American Journal of Dentofacial Orthopaedics*, 99, 298–309.

Parrington, F. R. (1934). On the cynodont genus *Galesaurus*, with a note on the functional significance of the changes in the evolution of the theriodont skull. *Annals and Magazine of Natural History*, 13, 38–67.

Pennycuick, C. J. (1967). The strength of the pigeon's wing bones in relation to their function. *Journal of Experimental Biology*, 46, 219–233.

Perry, C. C. and Lissner, H. R. (1962). *The Strain Gauge Primer* (2nd edition). New York: McGraw-Hill.

Peterson, O. A. (1914). The osteology of *Promerycochoerus*. *Annals of the Carnegie Museum*, 9, 149–219.

Piffault, C. and Duhamel, J. (1963). A propos des réactions subies par les condyles maxillaires a l'équilibre. *Revieu D'odonto-Stomatologie*, 21, 13–17.

Piveteau, J. (1961). *Traité de paléontologie*. VI 1 Paris: Masson et Cie.

Radinsky, L. B. (1981). Evolution of skull shape in carnivores. I. Representative modern carnivores. *Biological Journal of the Linnean Society*, 15, 369–388.

Radinsky, L. B. (1985). Patterns in the evolution of ungulate jaw shape. *American Zoologist*, 25, 303–314.

Ravosa, M. J. (1988). Browridge development in Cercopithecidae: A test of two models. *American Journal of Physical Anthropology*, 76, 535–555.

Ravosa, M. J. (1991a). Ontogenetic perspective on mechanical and nonmechanical models of primate circumorbital morphology. *American Journal of Physical Anthropology*, 85, 95–112.

Ravosa, M. J. (1991b). Interspecific perspective on mechanical and nonmechanical models of primate circumorbital morphology. *American Journal of Physical Anthropology*, 86, 369–396.

Ravosa, M. J. and Hylander, W. L. (1993). Functional significance of an ossified mandibular symphysis: A reply. *American Journal of Physical Anthropology*, 90, 509–512.

Ravosa, M. J., Hylander, W. L. and Kowalski, E. M. (2000). Masticatory stress, orbital orientation and the evolution of the primate postorbital bar. *Journal of Human Evolution*, 38, 667–693.

Rensberger, J. M. (1973). An occlusion model for mastication and dental wear in herbivorous mammals. *Journal of Paleontology*, 47, 515–528.

Rensberger, J. M. (1986). Early chewing mechanisms in mammalian herbivores. *Paleobiology*, 12, 474–494.

Rigler, L. and Mlinsek, B. (1968). Die Symphyse der Mandibula beim Rinde. *Anatomischer Anzeiger*, 122, 293–314.

Roberts, D. and Tattersall, I. (1974). Skull form and the mechanics of mandibular elevation in mammals. *American Museum Novitates*, 2536, 1–9.

Romer, A. S. (1956). *Osteology of the Reptiles*. Chicago: University of Chicago Press.

Romer, A. S. (1966). *Vertebrate Paleontology*. Chicago: University of Chicago Press.

Root-Bernstein, R. S. (1989). *Discovering*. Cambridge, Massachusetts: Harvard University Press.

Rosenberg, H. and Gans, C. (1977). Preliminary analysis of mastication in *Sphenodon punctatus*. *American Zoologist*, 17, 871, (Abstract).

Rosenberger, A. L. (1986). Platyrrhines, catarrhines and the anthropoid transition. In: *Major Topics in Primate and Human Evolution*. ed. B. A. Wood, L. Martin, and P. Andrews. Cambridge: Cambridge University Press, pp. 66–88.

Rosenzweig, M. L. (1966). Community structure in sympatric Carnivora. *Journal of Mammalogy*, 47, 602–612.

Ross, C. F. (1995). Muscular and osseous anatomy of the primate anterior temporal fossa and the functions of the postorbital septum. *American Journal of Physical Anthropology*, 98, 275–306.

Ross, C. F. and Hylander, W. L. (1996). *In vivo* and *in vitro* bone strain in the owl monkey circumorbital region and the function of the postorbital septum. *American Journal of Physical Anthropology*, 101, 183–215.

Rudwick, M. J. S. (1964). The inference of function from structure in fossils. *British Journal of the Philosophy of Science*, 15, 27–40.

Russell, M. D. (1985). The supraorbital torus: "A Most Remarkable Peculiarity". *Current Anthropology*, 26, 337–360.

Russell, M. D. (1986). In-vivo testing and refinement of the frame model of the craniofacial skeleton. *American Journal of Physical Anthropology*, 69, 259.

Ryan, A. S. (1979a). A preliminary scanning electron microscope examination of wear striation direction on primate teeth. *Journal of Dental Research*, 58, 525–530.

Ryan, A. S. (1979b). Wear striation direction on primate teeth: a scanning electron microscope examination. *American Journal of Physical Anthropology*, 50, 155–168.

Scapino, R. P. (1965). The third joint of the canine jaw. *Journal of Morphology*, 116, 23–50.

Scapino, R. P. (1972). Adaptive radiation of mammalian jaws. In: *Morphology of the Maxillo-mandibular Apparatus*. ed. G. H. Schumacher. IXth International Congress of Anatomists. Leipzig: G. Thieme, pp. 33–39.

Scapino, R. P. (1976a). The jaw symphysis in cats and bears. *American Journal of Physical Anthropology*, 44, 204 (Abstract).

Scapino, R. P. (1976b). Function of the digastric muscle in carnivores. *Journal of Morphology*, 50, 843–860.

Scapino, R. P. (1981). Morphological investigation into functions of the jaw symphysis in carnivorans. *Journal of Morphology*, 167, 339–375.

Schultz, C. B. and Falkenbach, C. H. (1940–1968). Contributions to the revision of the oreodonts (*Merycoidodontidae*), Nos. 1–8. *Bulletin of the American Museum of Natural History*, 77, 213–306 (1940); 79, 1–105 (1941); 88, 159–285 (1947); 93, 69–198 (1949); 95, 87–150 (1950); 105, 143–256 (1954); 109, 373–482 (1956); 139, 1–498 (1968).

Schumacher, G. H. (1973). The head muscles and hyolaryngeal skeleton of turtles and crocodilians. In: *Biology of the Reptilia. Vol. 4*. ed. C. Gans, and T.S. Parsons. New York: Academic Press, pp. 101–199.

Scott, W. B. (1937). *A History of Land Animals in the Western Hemisphere* (2nd edition). New York: Macmillan.

Shute, C. C. D. (1954). Function of the jaws in mammals. *Journal of Anatomy*, London, 88, 565 (Abstract).

Simpson, G. G., Roe, A. and Lewonton, R. C. (1960). *Quantitative Zoology* (revised edition). New York: Harcourt, Brace and World.

Smith, R. J. (1978). Mandibular biomechanics and temporomandibular joint function in primates. *American Journal of Physical Anthropology*, 49, 341–350.

Snodgrass, R. E. (1950). Comparative studies on the jaws of mandibulate arthropods. *Smithsonian Miscellanious Collections*, 116, 1–85.

Spencer, M. A. (1991). *Primate Masticatory System Configuration and Function: A Three-dimensional Biomechanical Approach*. Unpublished PhD proposal, pp. 1–147. Doctoral Program in Anthropological Sciences, State University of New York, Stony Brook.

Spencer, M. A. (1994). Changes in masticatory muscle activity with bite point position: an electromyographic test of biomechanical hypotheses. *American Journal of Physical Anthropology*, 18, 184. (Abstract)

Spencer, M. A. (1995). *Masticatory system configuration and diet in anthropoid primates*. PhD dissertation, State University of New York, Stony Brook.

Spencer, M. A. (1996). Functional constraints on masticatory system form in anthropoid primates. *American Journal of Physical Anthropology*, 22, 218 (Abstract).

Spencer, M. A. (1997). Force production in the primate masticatory system: electromyographic tests of biomechanical hypotheses. *Journal of Human Evolution*, 34, 25–54.

Spencer, M. A. (1998). Tooth root morphology and diet in primates. *American Journal of Physical Anthropology*, 107, 206 (Abstract).

Spencer, M. A. (1999). Constraints on masticatory system evolution in anthropoid primates. *American Journal of Physical Anthropology*, 108, 483–506.

Spencer, M. A. and Demes, B. (1993). Biomechanical analysis of masticatory system configuration in Neandertals and Inuits. *American Journal of Physical Anthropology*, 91, 1–20.

Stern, J. T., Jr. (1974). Computer modeling of gross muscle dynamics. *Journal of Biomechanics*, 7, 411–428.

Stöckmann, W. (1975). *Die Form der Mandibel Afrikanischer Bovidae (Mammalia) und ihre Beeinflussung durch die Ernährung*. PhD dissertation. Hamburg, Germany: University of Hamburg.

Stöckmann, W. (1979). Formenunterschiede der Mandibel Africanischer Bovidae (Mammalia) und ihre Beziehungen zur Zusammensetzung der Nahrung. *Zoologische Jahrbücher für Systematik, Oekologie und Geographie der Tiere*, 106, 344–373.

Storch, G. (1968). Functionsmorphologische Untersuchungen an der Kaumuskulatur und an Korrelierten Schädelstrukturen der Chiropteren. *Abhandlungen der Senckenbergischen Naturforschenden Gesellshaft*, 517, 1–92.

Tattersall, I. (1972). The functional significance of airorhynchy in *Megaladapis*. *Folia Primatologica*, 18, 20–26.

Teaford, M. F. (1985). Molar microwear and diet in the genus *Cebus*. *American Journal of Physical Anthropology*, 66, 363–370.

Thomason, J. J. (1991). Cranial strength in relation to estimated biting forces in some mammals. *Canadian Journal of Zoology*, 69, 2326–2333.

Thomason, J. J. and Russell, A. P. (1986). Mechanical factors in the evolution of the mammalian secondary palate: a theoretical analysis. *Journal of Morphology*, 189, 199–213.

Thompson, E. N., Biknevicius, A. R. and German, R. Z. (2003). Ontogeny of feeding function in the gray short-tailed opossum *Monodelphis domestica*: empirical support for the constrained model of jaw biomechanics. *Journal of Experimental Biology*, 206, 923–932.

Thorpe, M. R. (1937). The Merycoidodontidae an extinct group of ruminant mammals. *Memoirs of the Peabody Museum of Natural History*, III (4), 1–428.
Turnbull, W. D. (1970). Mammalian masticatory apparatus. *Fieldiana: Geology,* 18, 147–356.
Vizcaíno, S. F. and Bargo, M. S. (1998). The masticatory apparatus of the armadillo *Eutatus* (Mammalia, Cingulata) and some allied genera: paleobiology and evolution. *Paleobiology*, 24, 371–383.
Vizcaíno, S. F. and De Iuliis, G. (2003). Evidence for advanced carnivory in fossil armadillos (Mammalia: Xenarthra: Dasypodidae). *Paleobiology*, 29, 123–138.
Wainwright, S. A., Biggs, W. D., Currey, J. D., and Gosline, J. M. (1976). *Mechanical Design in Organisms.* London: Edward Arnold.
Walker, A. (1976). A 3-dimensional analysis of the mechanics of mastication. *American Journal of Physical Anthropology,* 44, 213 (Abstract).
Walker, A. (1978). Functional anatomy of oral tissues: mastication and deglutition. In: *Textbook of Oral Biology*. ed. J. H. Shaw, et al. Philadelphia: W.B. Saunders, pp. 277–296.
Walker, A. (1984). Mechanisms of honing in the male baboon canine. *American Journal of Physical Anthropology*, 65, 47–60.
Walker, W. F., Jr. (1970). *Vertebrate Dissection*. Philadelphia: Saunders.
Washburn, S. L. (1947). The relation of the temporal muscle to the form of the skull. *Anatomical Record*, 99, 239–248.
Weijs, W. A. (1980). Biomechanical models and the analysis of form: A study of the mammalian masticatory apparatus. *American Zoologist*, 20, 707–719.
Weijs, W. A. (1981). Mechanical loading of the human jaw joint during unilateral biting. *Acta Morphologica Neerlando-Scandinavica*, 19, 261–262.
Weijs, W. A. and Dantuma, R. (1981). Functional anatomy of the masticatory apparatus in the rabbit (*Oryctolagus cuniculus* L.). *Netherlands Journal of Zoology*, 31, 99–147.
Weijs, W. A. and de Jongh, H. J. (1977). Strain in mandibular alveolar bone during mastication in the rabbit. *Archives of Oral Biology*, 22, 667–675.
Weishampel, D. B. (1993). Beams and machines: modeling approaches to the analysis of skull form and function. In: *The Skull*. Vol. 3. ed. J. Hanken, and B. K. Hall. Chicago: University of Chicago Press, pp. 303–344.
Weishampel, D. B. (1995). Fossils, function, and phylogeny. In: *Functional Morphology in Vertebrate Paleontology*. ed. J. J. Thomason. Cambridge: Cambridge University Press, pp. 34–54.
Werdelin, L. (1986). Comparison of skull shape in marsupial and placental carnivores. *Australian Journal of Zoology*, 34, 109–117.
Werdelin, L. (1987). Jaw geometry and molar morphology in marsupial carnivores: Analysis of a constraint and its macroevolutionary consequences. *Paleobiology,* 13, 342–350.
Werdelin, L. (1988). Circumventing a constraint: the case of *Thylacoleo* (Marsupialia: Thylacoleonidae). *Australian Journal of Zoology*, 36, 365–571.
Werdelin, L. (1989). Constraint and adaptation in the bone-cracking canid *Osteoborus* (Mammalia: Canidae). *Paleobiology,* 15, 387–401.
Westneat, M. W. (2003). A biomechanical model for analysis of muscle force, power output and lower jaw motion in fishes. *Journal of Theoretical Biology*, 223, 269–281.
Whitmore, F. C., Jr. (1953). Cranial morphology of some Oligocene Artiodactyla. *U.S. Geological Survey Professional Papers,* 243-H, 117–159.
Willemse, M. C. A. (1950). The shifting of the molar row with regard to the orbit in *Equus* and *Giraffa*. *Zoologica Mededilingus, Leiden*, 30, 311–326.

Williams, G. (1956). The relationship between the length of the jaw and the length of the molar series in some eutherian mammals. *Proceedings of the Zoological Society, London*, 126, 51–64.

Wilson, J. A. (1974). Early Tertiary vertebrate faunas, Vieja Group and Buck Hill Group, Trans-Pecos Texas: *Protoceratidae, Camelidae, Hypertragulidae. Texas Memorial Museum Bulletin*, 23, 1–34.

Witmer, L. M. and Rose, K. D. (1991). Biomechanics of the jaw apparatus of the gigantic Eocene bird *Diatryma*: implications for diet and mode of life. *Paleobiology*, 17, 95–120.

Index

angular process, 4, 79, 80
antelopes, 41, 46, 74, 75, 77, 84, 85, 86, 88, 98
area above the curve, 22, 23
area below the curve, 19, 22
articular process, 3, 4

badger, 16
basic structure, 1
bears, 78, 92
Becht, 56, 101
braincase, 54, 65, 66, 67, 77, 79, 85, 86, 87, 89, 90, 93, 94
buttress, 27, 85, 89, 90, 93, 95, 99, 100

carnassial, 44
Carnivora, 45, 90
carnivores, 44, 45, 46, 49, 54, 61, 65, 75, 90, 92, 93, 94, 99
cat, 1, 92
chewing cycle, 21, 26, 78
components of force, 56, 57, 78, 88, 89, 90
compound distance, 67
compression, 15, 17, 25, 88, 90, 94, 95
contour map, 41
coronoid process, 3, 4, 70, 71, 101

dental arches, 40
dentine, 50, 51, 97, 98
diastema, 47, 48, 75, 76, 77, 78, 80, 93, 98
direction of travel, 50
disarticulated, 14, 33, 36, 50, 97
dislocated, 14, 33, 35, 59
dislocation, 17, 18, 25, 29, 33, 34, 35, 51
dissection, 24, 25, 52, 59, 72, 82, 96, 99
diversity, 1, 100
dogs, 46, 79, 90, 92

enamel, 50, 51, 97, 98
entelodonts, 1, 80
equilibrium, 5, 7, 8, 10, 11, 14, 16, 29, 31
evolutionary time, 45, 52, 58

facial vacuities, 89, 99
fossil, 1, 22, 79, 80, 89
four points, 52, 55, 67, 68, 69, 72

fulcrum, 5, 6, 7, 14, 16, 31
fusion, 27

gape, 11, 12, 16, 44, 81
glenoid, 1, 18, 48, 49, 54, 58, 59, 71, 72, 84, 104
Glyptodon, 79, 80
grand resultant, 28, 29, 30, 31, 32, 34, 35, 37, 40

helical tracts, 89, 90, 92, 93, 94
helices, 85, 88, 89, 90, 91, 92, 93, 94, 99
herbivores, 54
horses, 48, 49, 77, 78, 89, 104
hyenas, 92, 94, 100

incisors, 9, 18, 21, 27, 36, 46, 47, 77, 78
isodyne, 41, 42, 43, 51

jaw lengthening, 47, 52, 54, 76, 77, 78, 80
joint ligaments, 14, 15, 16, 17, 18, 25, 29, 33, 34, 35, 36

leading edge, 50, 97
lengthened jaw, 48, 75, 77, 93, 98

mechanical advantage, 11, 18, 19, 55, 96
metabolic, 61, 65, 66
minimum distance, 62, 64, 68, 72, 78
minks, 92
mongooses, 92
multiplying effect, 14, 15, 17, 18
muscle weights, 81, 82, 83
mustelids, 92, 93

natural design, 1, 3, 17, 95
non-biting side, 26, 27, 33, 35, 36, 39, 40, 43, 85, 86, 97
nutcracker, 14

occlusion, 2, 26, 27, 45, 49, 97
occlusal plane, 56, 57
orbital region, 68, 85, 87, 93, 99
otters, 92
output, 3, 7, 8, 9, 15, 21, 22, 40, 51

pigs, 70
pivot, 5, 45

point T, 65, 66, 67, 68, 69, 70, 72, 96
postorbital bar, 77, 85, 88, 89, 90, 93, 94, 98, 99
power stroke, 26, 56
premolars, 18, 36, 47, 48, 75, 77, 93
primates, 27, 46, 47, 70, 73, 89
projected lengths, 59
projection, 55

quotient, 73, 82, 83

rabbits, 75, 77, 78, 84, 90, 98
relative forces, 6, 49
reptiles, 34
ring shears, 45
rodents, 2, 41, 74, 75, 77, 78, 84, 90, 98
rostrum, 77, 85, 87, 90, 91, 94, 99

saber-toothed cats, 1, 54
safety factor, 33, 35, 43
scissors, 45, 49
selenodont artiodactyls, 47, 48, 49, 50, 77, 78, 85, 88, 89, 93, 98, 99
shear strains, 85, 88, 89, 93, 99
sum of the bite forces, 3, 19, 22, 23, 24, 25, 45, 46, 48, 51, 97
sum of the joint forces, 22, 23, 24
sum of the two distances, 66

temporalis, 71, 74
tensile, 14, 25, 33, 85, 88
tension, 15, 16, 25, 26, 27, 29, 88
tooth force, 7, 8, 61, 86
tooth wear, 50
torsional forces, 87, 88, 90, 92, 93, 99
trailing edge, 50, 98
triangle-of-support, 27, 28, 29, 31, 34, 43, 50
triangular plate, 27, 29, 30, 97
turning effect, 7
turtles, 34

unilateral biting, 85
unit vector, 4, 6, 10, 64, 65, 67, 68, 95
ursids, 92

viverrids, 92

wear facet, 50, 51, 97